BestMasters

Michael Lepper

Insights into the Adsorption Behavior of a Prototype Functional Molecule

A Scanning Tunneling Microscopy Study

Foreword by PD Dr. Hubertus Marbach and
Prof. Dr. Hans-Peter Steinrück

 Springer Spektrum

Michael Lepper
Erlangen-Nuremberg, Germany

OnlinePLUS material to this book can be available on
http://www.springer-imprint.de/978-3-658-11046-8

BestMasters
ISBN 978-3-658-11046-8   ISBN 978-3-658-11047-5 (eBook)
DOI 10.1007/978-3-658-11047-5

Library of Congress Control Number: 2015947132

Springer Spektrum

Printed on acid-free paper

Springer Spektrum is a brand of Springer Fachmedien Wiesbaden
Springer Fachmedien Wiesbaden is part of Springer Science+Business Media
(www.springer.com)

Foreword

With the invention of the scanning tunneling microscope (STM) in the early 80's of the last century, the direct real space observation of atoms and molecules on extended solid surfaces became possible. Due to its unparalleled capabilities, the STM quickly established itself as a powerful tool in fundamental sciences and is nowadays a standard instrument in laboratories all over the world. The investigation of large organic molecules on surfaces by STM has become a vivid research field with the vista to engineer functional devices. Specific interactions between absorbed molecules and/or with the underlying substrate often trigger peculiar adsorption behaviors like the self-assembly into long range ordered arrays. Due to their versatility, the molecules from the "porphyrin family" are considered as ideal building blocks for the generation of functional molecular devices: they combine a rigid planar framework as a structure-forming element with an intrinsic functionality, which is mainly determined by the coordinated metal center. The importance of porphyrins is highlighted by their omnipresence as main functional building blocks in nature - examples are iron porphyrin in heme or magnesium porphyrin in chlorophyll - but also due to their application in sensor and solar cell technology. The enormous potential of porphyrins for the fabrication of tailor-made functional molecular architectures on well-defined substrates has stimulated significant activities in fundamental research.

The Master thesis of Michael Lepper is a significant contribution to the ongoing competitive research in this area. Using STM in an ultra-high vacuum environment, he performed a very detailed microscopic study of a new porphyrin species, that is, Nickel-Tetraphenyltetrabenzoporphyrin (Ni-TPBP), adsorbed on a single crystalline copper surface. The understanding of the adsorption behavior of this functional molecule provides important information for the controlled preparation of molecular architectures. The applied „bottom-up" approach receives increasing attention in view of the limitations of classical „top-down" approaches at the molecular length scale. The study by Michael Lepper provides detailed insights in the very complex adsorption behavior of Ni-TPBP. He is able to image the molecules with submolecular resolution at room temperature and determines their surface arrangement with high accuracy. One of the most fascinating aspects is the coexistence of different, highly ordered phases. From the data, a convincing interpretation of the supramolecular arrangement and intramolecular conformation is deduced. One of the structures appears particularly promising, as it consists of Ni-TPBP molecules in two different conformations, which bear the potential for future molecular switches at room temperature.

Michael Lepper performed his Master thesis at the Chair of Physical Chemistry II of the University Erlangen-Nuremberg, which focuses on surface and interface science. Main research interests are (1) the development of new materials with novel electronic, geometric and chemical properties, (2) the investigation of elementary steps of surface reactions and (3) the development and construction of advanced scientific apparatus. The investigations aim at a fundamental physical and chemical understanding of all mechanisms and processes involved, at an atomic level, and at the identification and evaluation of new preparation routes. Of particular interest are detailed investigations of model surface reactions with particular emphasis on the influence of the surface structure on reactivity and reaction pathways. For these investigations a large variety of experimental methods is applied, including synchrotron radiation-based photoelectron spectroscopy, scanning electron and scanning tunneling microscopy, and molecular beam methods.

Hubertus Marbach and Hans-Peter Steinrück (supervisors)

Preface

I would like to thank all the people who supported and contributed to this master thesis and its publication, especially:

Prof. Dr. Hans-Peter Steinrück for the support and the possibility to do this thesis in his workgroup. PD Dr. Hubertus Marbach for his great support, input and time and of course the possibility to do this thesis in the STM group. Michael Stark for the introduction to STM and to the machine as well as for his help not only in the lab but in every way. Stefanie Ditze and Liang Zhang for their help in the lab and the fruitful discussions making the work in the lab and on my thesis more fun. Wolfgang Hieringer and his group for the contribution of the DFT gas phase calculations. Norbert Jux and his group for providing Ni-TPBP. H.-P. Bäumler, Bernd Kreß, Uwe Sauer and the workshop for support and technical expertise. Thanks also to the rest of the working group PCII for a great working atmosphere.

Finally I want to thank my parents and my brothers as well as my friends who supported me over all the years of studies and during the making of my master thesis.

Michael Lepper

Table of Contents

All figures can be accessed on www.springer.com under the author's name and the book title.

List of Abbreviations

AES   Auger Electron Spectroscopy

CCM   Constant Current Mode

CHM   Constant Height Mode

Co-TPP   Cobalt-Tetraphenylporphyrin

Co-TTBPP   Cobalt-Tetrakistertbutylphenylporphyrin

Cu-TPP   Copper (Cu)-Tetraphenylporphyrin

DFT   Density Functional Theory

fcc   face centered cubic

(L)DOS   (Local) Density Of States

LEED   Low Energy Electron Diffraction

LN   Liquid Nitrogen

M-TPP   Metallo-Tetraphenylporphyrin

Ni-TPBP   Nickel-Tetraphenyltetrabenzoporphyrin

QMS   Quadrupole Mass Spectrometry

RT   Room Temperature

STM   Scanning Tunneling Microscopy

TBP   Tetrabenzoporphyrin

TPP   Tetraphenylporphyrin

UHV   Ultra High Vacuum

2HTPP   free base (2H) Tetraphenylporphyrin

List of Abbreviations

1. Introduction

In modern technology there is an increasing demand for miniaturization of functional building blocks. Therefore engineering of systems on a molecular level by 'bottom up' approaches is desirable in contrast to classical 'top down' approaches like UV-lithography in semiconductor industry.[1] One promising way to realize the fabrication of functional structures from molecules is through self-assembly on well-defined surfaces.[2-5] In order to control the arrangement of the corresponding molecular building blocks on the surface it is essential to characterize and understand the adsorbate-adsorbate as well as the adsorbate-substrate interactions which determine the assembly processes. In this regard scanning tunneling microscopy (STM) has proven to be a very suitable method to determine the adsorption behavior of large organic molecules on solid surfaces[1, 3, 6] and partially also their intramolecular conformation.[7-9]

Porphyrins are certainly the most promising candidates as building blocks for the fabrication of functional molecular architectures. They offer a rigid but not too inflexible framework exhibiting a delocalized π-electron system and a central cavity which can host a metal center. The intrinsic functionality of a particular porphyrin species is determined by the nature of the central metal, the attached peripheral ligands as well as the particular interaction with the substrate in the case of surface confined molecules. Therefore their properties can be modified and tuned without changing the central porphyrin skeletal structure.

The versatile functionalities of porphyrins can be highlighted by their role as main functional building blocks in vital natural processes like cell respiration, detoxification of xenobiotics, oxygen transport, fatty acid oxidation and light harvesting.[10] In addition porphyrins have already become an integral part in numerous applications, e.g., in catalysis, optical sensors, electronic devices, supramolecular systems, photovoltaic cells, solar energy conversion and medicine because of their outstanding properties.[10]

The adsorption behavior of porphyrins has already been elaborately investigated with STM.[8-9, 11-13] However, an inspiring class of molecules can be obtained by combining the molecular framework of porphyrins and phthalocyanines.[14] Hence the name 'tetrabenzoporphyrins' was introduced to describe this hybrid class of molecules which feature the same constitution of the macrocycle as porphyrins, but are characterized by an additional benzene ring fused to the beta positions of each pyrrole ring. Due to their constitutional similarity to porphyrins tetrabenzoporphyrins exhibit related properties and also an extensive potential for applications. In contrast to porphyrins the class of tetrabenzoporphyrins is known to feature a red shifted Q-band in UV-Vis spectra. This optical behavior, coupled with the property to generate singlet oxygen, makes them suitable for photodynamic therapy.[10, 15-16] Tetrabenzoporphyrins also play an important role in the field of novel organic semiconductors applied for example in AM-OLEDs.[17-18] Furthermore tetrabenzoporphyrins are suitable for application in solution-processed, polycrystalline thin-film field-effect transistors.[19]

Obviously tetrabenzoporphyrins feature many useful properties but, contrary to porphyrins, no detailed characterization with STM has been reported up till now. Therefore the thesis at hand focuses on the investigation of the adsorption behavior of a tetrabenzoporphyrin, namely Ni(II)-tetraphenylbenzoporphyrin (Ni-TPBP), on a Cu(111) surface with STM under ultra-high vacuum (UHV) conditions at room temperature (RT).

The thesis starts with an introduction of the fundamentals of STM and the experimental techniques followed by the presentation of the investigated molecules. In the main part the experimental results, containing the coverage-dependent adsorption behavior of Ni-TPBP on Cu(111) at RT, are shown and discussed. The thesis finishes with the conclusions and a brief outlook.

2. Fundamentals

In the first part of the chapter fundamental theoretical and technical fundamentals of the used analytical instruments will be described.

In the second part of this chapter the substrate and the investigated organic molecules are introduced.

2.1. Scanning tunneling microscopy

The scanning tunneling microscope was invented by Gerd Binnig and Heinrich Rohrer in 1980. After the introduction into the scientific community in 1982 the outstanding capabilities of the method became clear since STM made it possible for the first time to analyze the topography and electronic structure of metal and semiconductor surfaces with atomic resolution in real space.[20] In 1983 the elucidation of the Si(111) 7x7 reconstruction proved to be the first scientific success of STM.[21] The importance and potential of scanning tunneling microscopy was underlined when Binnig and Rohrer received the Nobel Prize in Physics in 1986 for their invention. Nowadays STM remains an essential method for surface science. The operation principle of an STM is based on scanning a surface with a tip in close proximity and utilizing the resulting tunneling current for imaging and/or spectroscopy. The tunneling current is a pure quantum mechanical effect which will be explained in the next section.

2.1.1. The tunnel effect

In classic Newtonian dynamics a body cannot overcome or penetrate a potential wall that is higher than its energy (potential and kinetic). However, going from the macroscopic to the microscopic world Newton's laws of motion do not apply to certain processes. On the microscopic scale, for example for electrons, quantum mechanics has to be considered. In quantum mechanics the particle hitting the potential wall is actually able to penetrate this barrier with a certain probability. This pure quantum mechanical phenomenon is referred as tunneling effect.

In order to understand this effect some basic principles of quantum mechanics have to be introduced. Opposed to assigning a definite location and momentum to a body as in Newtonian dynamics, in quantum mechanics the state of a system like e.g. a photon or electron is defined by a wave function Ψ.[22] While the wave function itself has no physical meaning the square of its amplitude Ψ^2 at a certain point gives the probability of finding the particle at this position.

The description of quantum mechanical states and processes is done by combining wave functions and corresponding operators. The Schrödinger equation which is the most fundamental equation of quantum mechanics defines the total energy of a system E using the Hamilton operator $\hat{H}$.

$$\hat{H}\Psi = E\Psi$$

It is known that for the case of a 'particle in a box' the Schrödinger equation shows that the particle has some probability to be found outside the potential well.[23-24] For illustration and explanation of the tunnel effect a simple model system quite similar to the 'particle in a box' can be used (Figure 2.1.1-1). In this one-dimensional model the particle is travelling with a kinetic energy E_0 in a potential free region A and hits a potential barrier B of height $V_0 > E_0$ and width d. With a probability T the particle tunnels through the barrier into the potential free region C.

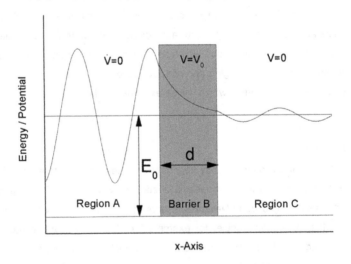

Figure 2.1.1-1 Scheme of a one–dimensional tunneling barrier.[24]

Using the Schrödinger equation it is quite easy to formulate the wave functions for the different regions in the model.

In region A the particle has the highest amplitude and thus maximum probability to be found. The wave function resembles an oscillation:

$$\Psi(x) = \Psi(0)\, e^{ix\sqrt{\frac{2m(E_0 - V)}{\hbar^2}}}$$

Within the barrier (region B) the amplitude of the wave function and thus the probabildity (Ψ^2) of the particle to be found at the given position are decreasing exponentially. In the following equation Δx equals the distance into the barrier:

$$\Psi(\Delta x) = \Psi(0)\, e^{-\Delta x\sqrt{\frac{2m(V_0 - E_0)}{\hbar^2}}}$$

In region C the maximum amplitude is reduced however the probability to find the particle in this region is not zero:

$$\Psi(x) = \Psi'(0)\, e^{ix\sqrt{\frac{2m(E_0 - V)}{\hbar^2}}}$$

The tunneling probability can be then put in the following equation:

$$T = \frac{4E_0(V_0 - E_o)}{V_0^2}\, e^{-\frac{2d}{\hbar}\sqrt{2m(V_0 - E_o)}}$$

Thus the probability depends on the energy E_0 and the mass m of the particle as well as the height V_0 and the width d of the potential barrier. It should be noted that even though the square of the wave function decreases, the wavelength resembling the energy of the particle remains the same after tunneling.

The principle of the simple model system shown above can be transferred to explain the tunneling process in STM (see figure 2.1.1-2). In this figure the sample represents the potential free region A while the other potential free region C is related to the tip. In principle a tunneling current can be measured without applying a bias voltage (eU). However in order to maintain the difference between Fermi energies and to address different orbitals for the tunneling process, generally a bias voltage is applied. It has to be noted that in this thesis the sign of the bias voltage denotes the polarity of the sample. Thus negative bias means tunneling from the sample to the tip and positive bias tunneling in the opposite direction.

Furthermore in the thesis at hand different adjectives describing the magnitude of bias voltage will be used. For definition these adjectives will relate to the absolute value of the voltage. For example bias voltages like ±200 mV will be referred as 'small', while ±1000 mV is considered 'large'.

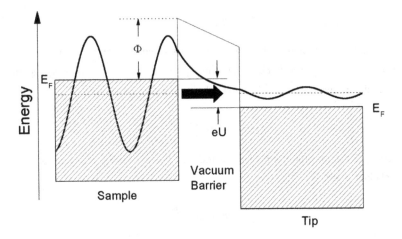

Figure 2.1.1-2 Schematic of the theoretical principle of STM. Modified from [25].

Coming back to figure 2.1.1-2, the barrier height is approximately the mean work function Φ of tip and sample while d is the width of the barrier. The grey shaded areas resemble the occupied states of sample and tip.

$$\Phi = \frac{1}{2}(\Phi_{sample} + \Phi_{tip})$$

For STM measurements a bias voltage (eU) is applied between tip and sample. The bias voltage (eU) yields an energy gradient by shifting the Fermi energies (E_F) of tip and sample. Thus in the case of figure 2.1.1-2 electrons from the occupied valence band of the sample can tunnel into the unoccupied conduction band of the tip.
This flow of electrons is the tunneling current I which depends exponentially on the sample to tip distance d.

$$I \propto U_{bias} \cdot \rho_s \cdot e^{-\frac{2d}{\hbar}\sqrt{2m\Phi}}$$

From this equation it can be derived that the tunneling current depends on the local density of states (LDOS) at the sample surface ρ_s and the tunneling bias U_{bias}. The LDOS is defined as number of electrons per unit volume per unit energy, at a certain point in space at a specific energy.[25]

The exponential dependence of the tunneling current on the sample to tip distance provides the high z-sensitivity of STM. Moreover the influence of the LDOS on the tunneling current supplies information about the electronic structure of the surface.

2.1.2. STM operation principle

The operation principle of a scanning tunneling microscope is illustrated in figure 2.1.2-1.

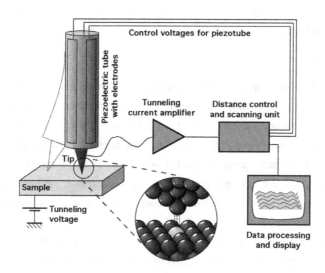

Figure 2.1.2-1 Schematic drawing of the operating principle of a scanning tunneling microscope.[26]

For an actual measurement a bias voltage between ±0.1 V and 3.0 V is applied and the sharp metallic tip is approached to the surface until a predetermined tunneling current is reached. The tip is then scanned over the surface at a distance in the nm regime and the corresponding tunneling current in the range of few pA to nA is measured. Control of the tip position is accomplished by applying specific voltages at a piezoelectric scanning tube on which the tip is mounted.

In STM two different modes to gain information from the tunneling current can be distinguished. Figure 2.1.2-2 illustrates these two modes.

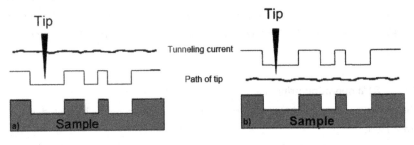

Figure   2.1.2-2   Comparison   of   the   two   STM   operation   modes:
a) constant height mode (CHM) and **b)** constant current mode (CCM).[27]

In the constant current mode (CCM) the sample is scanned at a constant tunneling current. A constant value of the tunneling current is maintained by a feedback loop that measures and counteracts the deviation of the actual tunneling current from the current setpoint. Thus the vertical tip position is adjusted such that the actual tunneling current remains at the given set point. The changes of the vertical tip position can then be translated into a color coded image by the software. However, the scan speed of this mode is limited by the restricted reaction time of the feedback loop.

Another way of scanning the tip across the surface is the constant-height mode (CHM) in which the z-position of the scanner is not changed and the current is recorded as a function of its lateral position on the surface. Therefore it is not necessary to implement a feedback loop. Thus the CHM can be applied with higher scan speeds than the CCM. In contrast a disadvantage of the CHM is the limitation to flat surfaces as surface defects and irregularities can cause a crash of the tip.

In the thesis at hand the STM was operated exclusively in constant current mode.

2.1.3. Tersoff-Hamann theory and related concepts

The one dimensional model for the tunnel effect explains only the basic principles of scanning tunneling microscopy. However, more complicated issues like the influence of the tip structure and shape cannot be understood with this simple model. In 1983 Tersoff and Hamann published a first model based on time-dependent perturbation theory enabling the calculation of the tunneling current considering the tip constitution.[28-29] The tunneling current was found to be proportional to the surface local density of states (LDOS) which are projected in constant current mode STM images. For this model the tip is assumed to be hemispherical with a spherical or s-like potential as depicted in figure 2.1.3-1. However soon after publication the limitations of the model became clear since it was not able to explain the atomic resolution of various substrates.

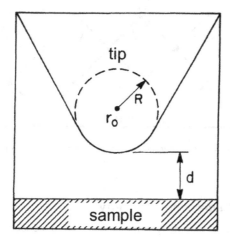

Figure 2.1.3-1 Spherical potential utilized by the Tersoff Hamann model to describe the tip state. Modified from [28].

Contemporary Baratoff proposed a model concerning the STM resolution of the Si(111) 7x7 reconstruction.[30] His model was based on a localized d_{z^2} orbital located at the apex of the applied tungsten tip.

The importance and existence of this d_{z^2} orbital was shown by 'first principle' calculations performed by Onishi et al. in 1989.[31] In 1991 Chen published the corresponding theory focusing on the interaction between localized states of tip and sample (see figure 2.1.3-2).[32] The significance of this theory is that it not only explains the origin of atomic resolution but also the principle of improving the tip. Application of pulses and crashing the tip are supposed to produce localized surface states through which tunneling takes place.

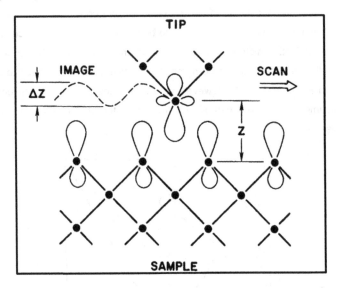

Figure 2.1.3-2 Illustration of the interactions between localized states of tip and sample.[32]

2.2. Quadrupole mass spectrometry

Today several mass spectrometry methods exist that dependent on different ionization processes and/or mass detection methods. Generally ionization and acceleration towards a mass filtering device take place in vacuum environment. The ionization process usually leads to characteristic fragments of the ionized molecule. This fragmentation can be used to determine the chemical composition and fragmentation behavior of the molecule.

After ionization and acceleration the particles of different mass and charge have to be separated according to their mass to charge ratio m/z and be detected. A typical mass spectrum is then an intensity spectrum of all possible mass to charge ratios of all fragments.

Figure 2.2-1 shows the working principle of a quadrupole mass filter. In a quadrupole mass spectrometer ion separation is accomplished by an oscillating electric quadrupole field. The quadrupole consists of four parallel rods where adjacent rods have opposite voltage polarity applied to them (see figure 2.2-1). The oscillating field is generated at the rods by a sum of constant DC voltage U and varying radio frequency ($U_{rf} \cos (\omega t)$), where ω = angular frequency of the radio frequency field. After the ionization in the source the ions are accelerated in a way that they travel parallel to the four rods. However for given DC and AC voltages, only ions of a certain mass to charge ratio pass through the quadrupole filter and all other ions are deflected from their original path. Thus by varying the oscillating frequency all possible m/z ratios can be obtained and a mass spectrum is received.

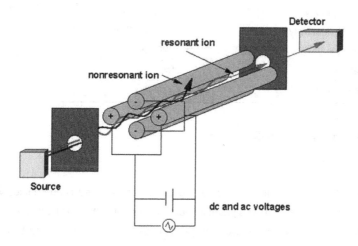

Figure 2.2-1 Setup of a quadrupole mass spectrometer.[33]

2.3. Substrate properties

The substrate used in the work at hand is a copper single crystal with a polished (111) surface. Table 2.3-1 shows an overview of several selected properties of copper.

properties	Cu
atomic number	29
relative atom mass	63,55
melting point	1084 °C
boiling point	2595 °C
electron configuration	[Ar] $3d^{10} 4s^1$
density	8,92 g/cm³
electronegativity	1,8 (Pauling)
electric conductivity (at 298K)	59,4 10^6 S/m
atomic radius	128 pm
crystal structure	fcc
lattice constant	3,61 Å

Table 2.3-1 Physical properties of copper.[34]

Figure 2.3-1 a) shows the unit cell of the cubic face centered copper lattice with a lattice constant of 3.61 Å. The green planes in a) resemble the (111) surface.

Figure 2.3-1 b) presents a topview of the (111) surface from which it can be derived that Cu (111) is hexagonal close packed. The planes are orientated perpendicular to the space diagonal of the fcc lattice. Within one layer the Bravais lattice is hexagonal with a lattice constant of 2.55 Å. As it is known for hexagonal close packed lattices the layers follow an ABC stacking where every fourth layer is congruent to the first. The atoms of one layer fill the vacant spaces of the respective upper layer leading to dense packing. In figure 2.3-1 b) the arrows are used to emphasize the three-fold high symmetry axes of Cu (111). Similar arrows will be used in later chapters to indicate the substrate orientation.

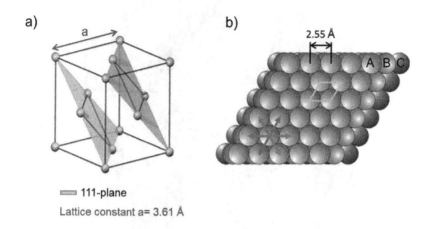

a)

a

111-plane
Lattice constant a= 3.61 Å

b)

2.55 Å

A B C

Figure 2.3-1 a) fcc unit cell with green planes representing a (111) cut.[35] **b)** topview of the hexagonal (111) surface. The orange arrows indicate the three-fold high symmetry axes. The hexagonal unit cell is symbolized by the green rectangle. Modified from [36].

In figure 2.3-2 a scanning tunneling micrograph of a freshly prepared clean Cu(111) surface is presented. Cleaning of the surface was performed with the procedure described later on in chapter 3.4.1.

The steps on the substrate have no defined form and show no systematic arrangement to each other. However considering the fcc lattice and the Cu(111) surface, it would be expected from theory that the angles between the steps should be multiples of 120 degrees. This observation is characteristic for a copper surface and can be explained by the low diffusion barrier of copper atoms on Cu(111).[37-38] For the theoretical quite similar Ag(111) surface this phenomenon is not observed and the steps align along the predominant directions of the crystal.

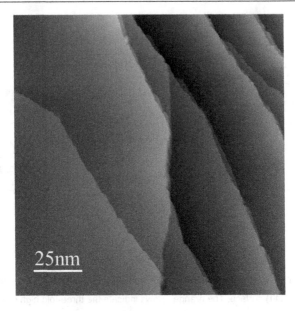

Figure 2.3-2 Scanning tunneling micrograph of a clean Cu(111) surface. (U_{bias} = -1 V, I_{set} = 29.1 pA)

2.4. The porphyrins

As already mentioned before, one important field in surface science is to investigate structures on the nano-scale fabricated by self-assembly of supramolecular architectures from functional molecules.[2-5]

One important class of molecules in that regard are porphyrins. They exhibit a rigid framework, can be synthesized in large variety and bind different metal atoms. These facilities of porphyrins are of importance because their chemical reactivity and selectivity are governed by their structure and the actual complexed central metal atom. Thus the adsorption behavior can be influenced by modifying the porphyrin structure without changing the central porphyrin macrocycle.

2.4.1. Main structure

The reoccurring feature in all porphyrins is the central framework which is illustrated in figure 2.4.1-1.

Generally a porphyrin is build up by four pyrrole rings that are connected by methin bridges in alpha position. These connections result in a macrocycle that contains 22 π-electrons.[39] Out of these, 18 π-electrons can form a delocalized π-system. Since the system fulfills Hückel's 4n+2 π-electron rule it is assumed aromatic which explains the color of porphyrins.[40]

The exemplary central framework depicted in figure 2.4.1-1 shows a free base porphyrin. However, it is also possible that a metal atom is complexed in the center of the macrocycle which will be explained later in more detail.

Furthermore there are two more possible substitution positions in the central part of a porphyrin. One is at the carbon atoms of the spacing methine bridges denominated 'meso' illustrated by the substituents R_2. The other one is directly in β-position at the pyrrole rings represented by R_1. So the macrocycle offers, besides the central cavity, four meso-positions and eight β-positions for substitution which enables a large variety.

In the next chapters the two different porphyrins which were examined in the work at hand will be introduced.

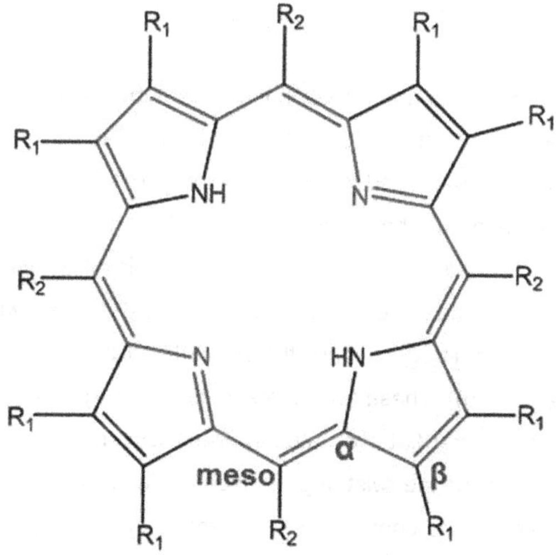

Figure 2.4.1-1 Sketch of the characteristic porphyrin framework. The red colored bonds indicate the delocalized π-electron system.

2.4.2. 2HTPP

One of the two molecules used in this work is 2H-5,10,15,20-tetraphenylporphyrin (2HTPP). Figure 2.4.2-1 shows a ball and stick model as well as a space filling model of 2HTPP. It becomes clear that 2HTPP is one of the most basic porphyrins since it is a free base porphyrin that only features four phenyl substituents in the meso-positions. Over the last years 2HTPP has become a well examined system that has been part of numerous STM-studies.

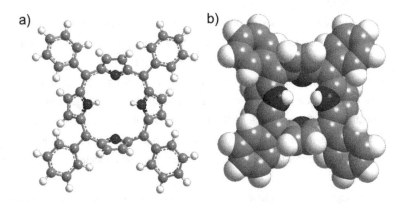

Figure 2.4.2-1 a) ball and stick and **b)** space filling model of 2HTPP.

Concerning the intramolecular conformation it is known that conformational adaptions of 2HTPP are observed for different substrates[41] Diller et al. showed that 2HTPP on Cu(111) exhibits a saddle-shaped conformation where the iminic nitrogen atoms point towards the surface and the pyrrolic nitrogen atoms point upwards.[42] This behavior is attributed to an attractive interaction of the iminic nitrogens with the surface copper atoms.[41] The inclination out of the macrocycle, described with the tilt angle, is α_{im} = -60°for the iminic nitrogens while the tilt angle for the aminic nitrogens is α_{am} = 40° (see figure 2.4.2-2 b)). These angles resemble a strong distortion of the macrocycle which leads to a rather flat orientation of the phenyl rings as compared to e.g. 2HTPP on Ag(111).[43] Thus the twist angle, which by definition describes the rotation of a certain group around its connection to the macrocycle, constitutes α_{ph} = 20°for 2HTPP on Cu(111). The flat orientation of the phenyl rings brings the macrocycle of 2HTPP closer to the surface and thus an enhanced interaction can be established (see lower part of figure 2.4.2-3 b).

Figure 2.4.2-2 a) shows the typical STM contrast found for 2HTPP on Cu(111) which is dominated by two longish protrusions. In figure 2.4.2-2 b) the parts of 2HTPP that can be seen in the scanning tunneling micrograph, namely the two pyrrole rings that are bent upwards and the phenyl rings, are colored yellow.

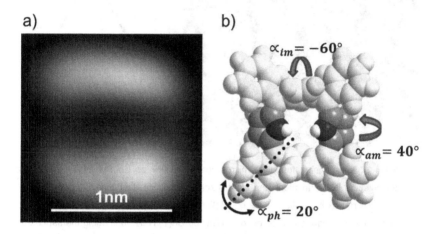

Figure 2.4.2-2 a) typical STM contrast found for 2HTPP on Cu(111) dominated by two longish protrusions. (U_{bias} = -0.77 V, I_{set} = 26.4 pA) **b)** space filling model of 2HTPP showing the twist and tilt angles. The parts of the structure which determine the observed STM contrast are marked yellow. Modified from [44].

Investigation of the adsorption behavior of 2HTPP on Cu(111) at RT leads to the observation of isolated molecules on the surface (see figure 2.4.2-3 a)).[13, 41] This is in contrast to the island formation found for 2HTPP on Ag(111).[45] However for 2HTPP on Cu(111), the dominating molecule-substrate interactions are obviously prevailing over the molecule-molecule interactions.[41]

Moreover it is observed that the isolated 2HTPP molecules are mobile on the surface and slowly move like a 'train on rails' along one of the three densely packed substrate directions at RT (see figure 2.4.2-3 a) and 2.4.2-3 b)). This movement is attributed to a high site selectivity of the adsorbate-substrate bond and was closely investigated by Buchner et al.[46] Thus 2HTPP can be used as a tracer for the three fold high symmetry directions of the Cu(111) substrate.

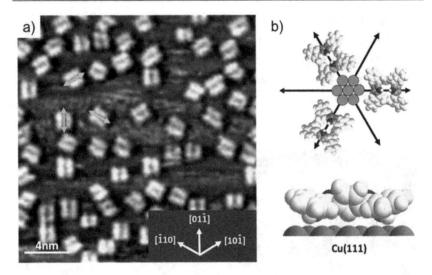

Figure 2.4.2-3 a) scanning tunneling micrograph of 2HTPP on Cu(111) showing isolated molecules which align along the close packed substrate directions. (U_{bias} = -0.77 V, I_{set} = 26.4 pA) **b)** Illustration of the orientation on the surface and the movement directions. The lower part is a sideview of 2HTPP on Cu(111) demonstrating the flat conformation of the molecule. Modified from [44].

Another feature of 2HTPP, that is connected to its strong molecular-substrate interaction with Cu(111) and should be briefly mentioned, is the so-called self metalation. In-situ metalation of 2HTPP on Cu(111) was investigated after annealing to 400 K for different time periods by STM as well as XPS measurements.[42, 47]

2.4.3. Ni-TPBP

An interesting class of porphyrin derivatives can be obtained by somewhat combining the molecular framework of porphyrins and phthalocyanines.[14] This hybrid class is called tetrabenzoporphyrins (TBP). In contrast to porphyrins, tetrabenzoporphyrins have an additional benzene fused to the beta positions of each pyrrole ring (see figure 2.4.3-1). Tetrabenzoporphyrins can be applied for example in photodynamic therapy but also in organic semiconductors applications.[10, 15]

The main topic of the thesis at hand was the investigation of Nickel-5,10,15,20-tetraphenyltetrabenzoporphyrin (Ni-TPBP) on Cu(111) with STM. Figure 2.4.3-1 illustrates the structure of Ni-TPBP. Similar to 2HTPP the meso-positions of the tetrabenzoporphyrin macrocycle bind the phenyl groups. In contrast to 2HTPP for Ni-TPBP the center of the macrocycle is occupied by a nickel atom. The metal atom is four fold coordinated with the nitrogen atoms of the pyrrole groups.

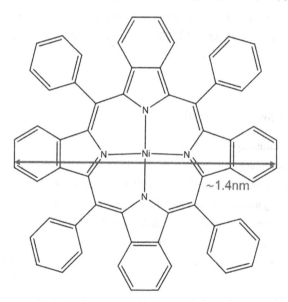

Figure 2.4.3-1 Chemical structure of Nickel- 5,10,15,20-tetraphenyltetrabenzoporphyrin (Ni-TPBP).

3. Experimental

All experiments in the thesis at hand were performed in a two chamber UHV system with a base pressure in the low 10^{-10} mbar regime. The system is divided in two main chambers: the STM chamber and the preparation chamber. Within the STM chamber the scanning tunneling microscope (RHK UHV VT RTM 300) is housed for the actual measurements. In the other chamber various sample preparation, characterization and manipulation tools are situated. The two chambers are linked by a VAT gate valve (CF150) which is usually only opened for transferring the sample. The sample transfer between the chambers is realized by a linear transfer rod and wobble sticks. Furthermore a fast entry lock is attached to the STM chamber enabling a quick sample exchange. Several pumps are used to generate and maintain the UHV in the system. The whole apparatus is carried by a steel frame which is placed on three laminar flow stabilizers to reduce the effect of low frequency vibrations (3-30 Hz) on the measurements. In the time in which this master thesis was prepared one of the three laminar flow stabilizers (Newport I-2000 series) was replaced by a Newport S-2000 series model.

3.1. The preparation chamber

3.1.1. The vacuum system

To establish and maintain the ultra high vacuum in the preparation chamber a three-way pumping layout is used. Monitoring of the vacuum conditions is done by a standard VG ion gauge connected to a VG IGC 26 control unit.

A turbo molecular pump (Balzers TMU 260) is attached to the system via a large gate valve to pump the system in cases of higher gas loads for example during Ar^{+}-sputtering or bake-out. The necessary rough vacuum ($\sim 10^{-3}$ mbar) is produced by a rotary vane pump.

During STM measurements the system is pumped by an ion getter pump (Varian Diode 500 l/s) which is sufficient to preserve the pressure in the low 10^{-10} mbar regime. The construction and working principle of the pump ensure it to be vibration free and thus enable operation during STM measurements without interference.

The last part of the three way pumping layout is titanium sublimation pump (TSP).

Titanium sublimation pumps are run with a certain time period coordinated by pro-grammable timer switches to maintain the low pressure regime in the chamber.

3.1.2. Instrumental setup

In the preparation chamber various instruments for preparing and characterizing the sample are situated.

In order to clean the surface of the sample an ion gun for Ar^+-sputtering (Specs IQE 11/35) was used. An external gas bottle ensured the supply of high purity argon gas to the ion gun via a gas inlet.

For depositing material on the crystal different evaporators are attached to the cham-ber. A water cooled Focus EFM 3 electron beam evaporator was used for metal evaporation. Another thermal evaporator was used for deposition of Tetraphenylpor-phyrin (2HTPP) which acts as a marker for copper substrate orientation in the case of the thesis at hand (see chapter 2.4.2). Deposition of Ni-TPBP was done by a home built Knudsen cell evaporator. Within the evaporator the organic substance is located in a glass crucible and is heated up by a tungsten filament to its sublimation tempera-ture. Since the whole evaporator is retractable and able to be sealed off from the preparation chamber by a UHV gate valve a change of the evaporant without break-ing the vacuum in the preparation chamber is possible.

Determination of the sublimation temperature of Ni-TPBP was accomplished by in-creasing the temperature of the evaporator stepwise and detecting the signal of the organic material at the corresponding mass with a quadrupole mass spectrometer (QMS). The QMS used in this UHV setup is a Pfeiffer QMG 700 system allowing the detection of molecules with a mass-to-charge ratio up to 2048.

In addition a four-grid LEED/AES optics (low energy diffraction/ Auger electron spec-troscopy) is attached to the chamber for initial characterization of the sample. How-ever, this was not used in the work at hand.

3.1.3. The manipulator

All the preparation and analysis tools mentioned above are mounted at different positions in the preparation chamber. Therefore a manipulator which allows independent linear movement in the x, y and z directions as well as rotation along the z-axis is needed for sample positioning relative to each tool.

Moreover the construction of the manipulator enables cooling and heating of the sample. Sample heating for temperatures up to 600 K was carried out by radiative heating from a retractable tungsten filament. Higher temperatures can be achieved with electron bombardment by using emitted electrons from the aforementioned filament in combination with high voltage (~600 V). Thereby temperatures over 1000 K can be reached.

The manipulator is equipped with a tube connected to a heat exchanger which is coupled to the sample holder by the sample mount made from massive copper. By feeding a cooling medium like liquid nitrogen through the tube cooling of the sample to temperatures as low as 100 K can be obtained.

Monitoring and controlling the sample temperature was done by a Eurotherm 818 temperature control unit. The controller is connected to the power source of the heating filament via an external control link and to the sample via the sample's integrated thermocouple. This makes it possible to apply defined temperature ramps to the sample or to maintain a constant temperature during an experiment.

3.2. The scanning tunneling microscope

3.2.1. The STM chamber

The pumping scheme for the STM chamber is quite similar to the one of the preparation chamber. An ion getter pump (Varian VacIon Plus 500) is used for maintaining the vacuum during an experiment. The pump setup is completed by a turbo pump (Balzers TPU 160), which is mainly used during bake-out and pumping the gas dosage system, and a titan sublimation pump (Varian 916-0017). The turbo pump is coupled to a second rough vacuum system similar to the one described in section 3.1.1. As in the preparation chamber the pressure in the STM chamber is also monitored by a VG ion gauge.

Besides the pump setup the STM chamber is equipped with a RHK 'Beetle' type 300 VT UHV STM system. The main components of the STM system are the sample stage on which the sample holder is placed for STM measurements, the scan head and the STM tip (see Figure 3.2.1-1). Furthermore a gas inlet for reactive gases is attached to the chamber.

The STM chamber houses a sample storage elevator for either six tip or sample holders. With a wobble stick and a transfer rod it is possible to move or transfer the tip or sample within the system.

Moreover the STM chamber is connected to a fast entry lock to insert new samples without breaking the vacuum of the whole system.

The layout of the STM chamber is depicted in figure 3.2.1-1.

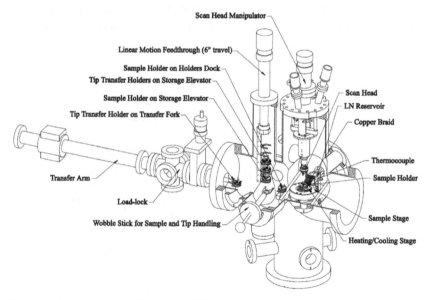

Figure 3.2.1-1 Setup of the STM chamber.[48]

3.2.2. The sample stage

The sample stage mount resembles the central part of the STM chamber (Figure 3.2.2-1). The 'U'-shaped fork part of the sample stage is used to insert the sample holder for STM measurements. Viton loops isolate the sample stage from the chamber in order to reduce vibrations which effect and disturb image acquisition. The sample stage also incorporates the heating/cooling stage as well as a copper braid connection to the cryostat (LN reservoir, see figure 3.2.2-1). Furthermore the stage holds electrical connections like the thermocouple connectors and bias voltage feed-throughs.

The design of the sample stage is in a way that heating and cooling of the sample is possible. Heating of the sample is accomplished by thermal radiation or electron bombardment using a filament which can be raised up into the bottom cavity of the sample holder. To cool the sample a cooling medium such as liquid or gaseous nitrogen is used. In the thesis at hand gaseous nitrogen was used because liquid nitrogen causes vibration.

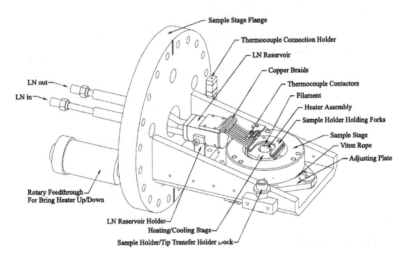

Figure 3.2.2-1 Setup of the sample stage.[48]

3.2.3. The scan head

The schematics of the 'Beetle' type STM scan head are depicted in figure 3.2.3-1.

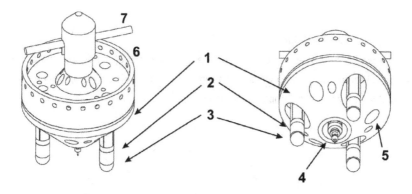

Figure 3.2.3-1 Scan head setup: (1) main body, (2) piezo tube leg, (3) sapphire ball, (4) tip holder, (5) notches for positioning tip transfer holder, (6) 'Mexican hat', (7) transverse pin. Modified from [48].

The main part of the scan head consists of a cone shaped metallic body (1). This metallic body is supported by the three piezo tube legs (2) which end in scratch resistant sapphire balls (3). Movement of the scan head is accomplished by the design of the piezo legs. Their outside is covered with four electrodes arranged along the tube axis. Inside the tubes a reference electrode is present. By applying a voltage at the inner and outer electrodes the tubes are bent according to the piezoelectric effect and linear x-y-movements in a slip-stick fashion can be realized.

In the center of the cone shaped metallic body another piezo tube is placed which ends in an exchangeable tip holder and is used for scanning. In order to correctly position the tip transfer holder used for the exchange of the STM tip the notches (5) between the piezo tube legs are used. The electrical connections for the piezo elements in form of thin Kapton isolated gold wires are attached to the scan head through holes in the upper part of the assembly (6). This part is often referred as "Mexican hat" due to its peculiar shape. With the scan head manipulator and the hanger assembly (7) the scan head can be lifted and lowered by hand. The transverse pin of the hanger assembly is inserted into a V-shaped window in the scan head manipulator. Through this design the scan head manipulator does not have any rigid mechanical connection to the scan head once it is lowered down onto the sample holder. This prevents vibration transmission to the scanning part of the setup.

3.2.4. The sample holder

The sample holder does not only house the sample but is of importance because of its helical ramp design onto which the scan head is lowered during operation. In figure 3.2.4-1 a schematic sample holder is shown.

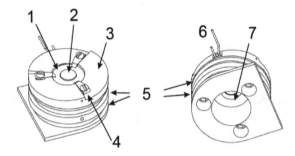

Figure 3.2.4-1 Sample holder setup: (1) sapphire washer, (2) sample, (3) helical ramp, (4) tungsten clamps, (5) grooved copper body, (6) thermocouple contacts, (7) cavity for heating filament. Modified from [48].

The main copper body of the sample holder includes two holding grooves (5). These grooves ensure firm hold for sample transport as well as placing the sample in the sample stage or in the preparation chamber manipulator. The U-shaped bottom part of the sample holder arranges the holder for equal alignment while entering the manipulator or sample stage. A central cavity in the sub section of the sample holder makes it possible to insert a heating filament (7).

As already mentioned the top part of the sample holder is built up by a three-segment helical ramp made of molybdenum (3). This segment acts as a platform for the three piezo legs of the scan head allowing for movement of the scan head on this platform either in an arbitrary x-y direction or in a turn motion around the center axis. For the coarse tip approach the turn motion is used because combined with the slight slope of the ramps it decreases the tip-sample distance.

In the central dimple of the ramp section the hat shaped sample (2) is located. Usually the sample is metal single crystal which is held by three tungsten clamps (4) between two sapphire washers (1) to avoid electrical contact to the sample holder. The final parts of the sample holder are the thermocouple contacts (6). The thermocouple itself is clamped between the sample and the sapphire washer to ensure good thermal contact to the sample holder.

3.2.5. The STM tip

The STM tip used in the thesis at hand is a Pt/Ir wire tip which is mechanically stable and suitable for STM at RT. Experience in our group has shown that STM tips can be prepared by clamping one end of a thin Pt/Ir wire (diameter: 0.2 or 0.25 mm) while angularly cutting the other end off.

However, the tip usually has to be manipulated before or during an experiment to achieve high quality scanning tunneling micrographs. One way of manipulating and optimizing the tip is by applying appropriate voltage pulses (<10 ms). Pulses of less than 1 V stabilize the tip through adding or removing adatoms to or from the tip. On the other hand high pulses (up to 10 V) have been proposed to enable a local rearrangement of tip atoms.[32]

If pulsing does not lead to a satisfying improvement of the micrographs a crash of the tip into the sample surface can be performed to completely reshape the tip.[48] A tip crash is induced by extending the scan piezo until a direct current is measured. Afterwards it might be necessary to apply several pulses to optimize the tip and achieve good resolution.

3.3. Electronic equipment

A scanning tunneling microscope also requires several electronic devices for operation.

Since only very low tunneling currents in STM occur it is necessary to pre-amplify the currents in order to measure them. The pre-amplifier used for this STM is a FEMTO DLPCA-200 variable-gain low-noise current amplifier with a maximum amplification factor of 10^9 V/A in low pass mode. It is placed as direct as possible onto the electrical vacuum feed-through of the chamber to minimize the signal path length and thus the possibility of picking up noise in the signal wire. The pre-amplifier is suitable for our setup because it allows tunneling currents in the 10 to 100 pA regime which fits the conditions needed for investigation of organic molecules on surfaces.

After transformation of the tunneling current into a DC voltage in the pre-amplifier the signal is fed into the main control electronics. During the preparation of the master thesis the electronic control box was upgraded from the SPM100 to the SPM200 (RHK). The SPM200 device is connected to a PC running the data acquisition software via a digital interface.

3.4. Procedures

3.4.1. Surface preparation

The sample used in this thesis is a copper single crystal (>99.999% purity, Mateck). Cu crystallizes in a face-centered cubic (fcc) lattice as already stated in chapter 2.2. The surface of the crystal is orientated along a (111)-plane and exhibits no spontaneous reconstruction.

Since the crystal had already been in use in the working group only minor cleaning steps between the experiments were needed. From prior experiences it is know that the following cleaning process is sufficient for the surface: In the first step the sample was sputtered at RT with Ar^+ ions ($E = 500$ eV) for about 90 minutes at a pressure of 5×10^{-5} mbar. Secondly, the sample was heated to 850 K through electron bombardment at a defined rate of 1 K/s ($E = 600$eV), the temperature was held at 850 K for 10 minutes. Afterwards the crystal was allowed to cool down to RT at a defined rate of 0.33 K/s.

However, during this heating step of the sample the manipulator was cooled to about 170 K using liquid nitrogen. This cooling step reduces desorption of residual gases from the hot manipulator which otherwise would lead to unwanted adsorption on the surface.

3.4.2. Organic layer deposition

For the organic molecules used in this thesis two different evaporators were used to prepare thin layers. In order to evaporate in an efficient way onto the sample the crystal was placed in front of the respective evaporator.

The organic species applied to the sample in the thesis at hand are Nickel-tetraphenyltetrabenzoporphyrin (Ni-TPBP) which was synthesized in Norbert Jux's group and Tetraphenylporphyrin (2HTPP) which was purchased from Porphrin Systems (specified purity >98%).

Ni-TPBP was evaporated and applied to the sample by the home-built Knudsen cell at sublimation temperature of 690 K. The sublimation temperature was determined with the procedure explained in chapter 3.1.2. For 2HTPP another evaporator and a sublimation temperature of 300 °C were used. Before deposition the evaporators were held at the respective sublimation temperatures for about 15 minutes to ensure a constant evaporation rate.

To prepare the layers the appropriate amount of material must be evaporated. However this approach is difficult because the flux of the evaporated material is dependent on the filling level of the evaporator and the evaporation temperature. By successively depositing the material in small amounts and monitoring the coverage after the different steps the desired sub monolayer coverage of the substrate was obtained.

3.4.3. Data acquisition and processing

Acquisition of the STM data was done using the XPMPro Software (version 2.1.0.5 and 2.1.0.6 RHK Technology).

For processing the images the WSxM 5.0 Develop Software is used.[49] The program supports basic manipulations such as background subtracts, smoothing and recoloring as well as more elaborate techniques like Fourier transformation and several filtering procedures.

Furthermore it offers tools to measure lattice parameters and height profiles of the surface. Concerning visualization the acquired STM data can be processed to 2D or 3D images.

In the thesis at hand the standard procedure for image processing with WSxM starts with the function 'plane' which resembles a plane fit. The 'linear flatten' function was used to correct inconsistency in the topography parallel to the scanning direction. Background noise can be reduced with filter functions like 'smooth' and 'gaussian smooth'. Another filter function is a 2D fourier filter which is used to reduce the contribution of high frequency vibration to the processed images. The contrast can be adjusted with the 'equalize' function. Moreover the image might also be recolored with the 'palette settings'.

Another important feature of WSxM is the 'movie' function which makes it possible to combines consecutive scanning tunneling micrographs to resemble a movie. The influence of thermal drift can be reduced while imaging the surface with a 'drift correction' function in XPMPro. STM-movies play an important role in analyzing dynamic processes on the surface.

4. Adsorption behavior of Ni-TPBP on Cu(111)

In order to design self-assembled supramolecular architectures with specific functional properties it is important to understand the adsorption behavior of molecular building blocks on well-defined substrates. One very versatile class of adsorbates in this regard are porphyrins which offer not only a rigid molecular framework but also ligand functionality to bind metals.

The main part of the thesis at hand is based on the STM investigation of the coverage-dependent supramolecular arrangement of Ni(II)-tetraphenylbenzoporphyrin (Ni-TPBP) on Cu(111) at room-temperature. By subsequent deposition the coverage of Ni-TPBP on Cu(111) was increased from low to medium and finally high coverage. Deposition of the organic molecules was performed as stated in section 3.4.2.

In the following sections, observations made at the different coverages will be described and interpreted.

4.1. Low coverage

Figure 4.1-1 a) depicts a representative scanning tunneling micrograph recorded at low coverage. The white square marks the area which can be seen in figure 4.1-1 b). At the step edges of the surface 'dot shaped' features, indicated by orange arrows, were observed. The dots arrange in chains appearing similar to a 'pearl necklace'. On the terraces no such dots were found. In fact stripy features were observed which coincide with the scan direction. For several of the chemically related M-TPPs, e.g. Co-TPP, a quite similar low coverage adsorption behavior featuring dots at the step edge and stripes on the terraces has been reported.[36, 43]

Since no individual molecules were detected on the terraces, it can be assumed that the interaction of Ni-TPBP molecules and the surface is weak. In terms of surface diffusion this means that the thermal energy (kT) of the molecules is higher than the migration barrier between adjacent sites (E_m).[50] The molecules are freely diffusing in a so-called 2D gas phase.[51] In this 2D gas phase the diffusion speed of the molecules is very high compared to the STM recording speed. Thus the recorded stripes resemble molecules that diffuse fast under the STM tip.

Considering the observation that the 'dot shaped' features occurred exclusively at the step edges, it is suggested in literature that the features correspond to adsorbed molecule.[43] One explanation for this adsorption behavior is that the step edges resemble energetically favored adsorption sites.[52-53]

To support this assumption for the examined case of Ni-TPBP on Cu(111) the distances of neighboring 'dots' were investigated. The mean distance between neighboring protrusions yields 1.36±0.12 nm. This value is in good agreement with the dimensions expected for one Ni-TPBP molecule (stated in section 2.4.3). Hence, the
measurements possibly indicate that two of the observed protrusions accord to one
Ni-TPBP molecule. In figure 4.1-1 b) the approach of two protrusions per molecule is
illustrated by an orange dumbbell drawing. The molecular appearance of Ni-TPBP
will be discussed in more detail in section 4.2.

Thus all observed molecules were located at the steps meaning that their diffusion
length at RT must be at least as long as the width of the corresponding terraces.[43]

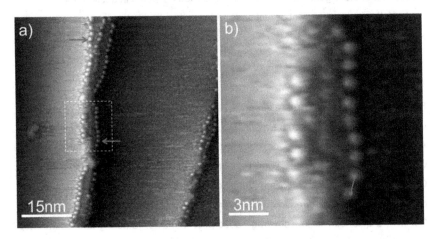

Figure 4.1-1 a) Scanning tunneling micrograph recorded at low coverage. (U_{bias} = +0.99 V,
I_{set} = 29.6 pA) The arrows indicate 'dot shaped' features attributed to step decoration. On
the surface stripy features were observed caused by fast diffusing molecules of the 2D
gas phase. The dashed-line rectangle marks the area presented in **b)** illustrating the
'pearl necklace' resemblance. The dumbbell drawing visualizes the approach of two protrusions per Ni-TPBP molecule.

4.2. Medium coverage

By successive deposition the coverage was increased to medium coverage in order to promote island formation of Ni-TPBP. Interestingly, it was possible to monitor three different coexisting molecular appearances of the Ni-TPBP islands. Figure 4.2-1 presents a scanning tunneling micrograph where the three different supramolecular arrangements coexisted.

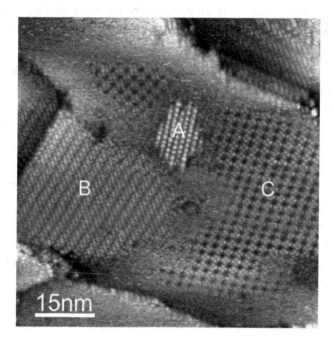

Figure 4.2-1 Scanning tunneling micrograph recorded at medium coverage. Three different island species were labeled: (A) marks the 'hexagonal' structured island, (B) the 'herringbone' ordered island and (C) the 'cross' structured island. (U_{bias} = +1.21 V, I_{set} = 28.8 pA)

In order to describe the islands more accurately figure 4.2-2 presents a close up of the three different island species. The first island type, marked with an 'A' in figure 4.2-1 and 4.2-2, was build up by bright protrusions that arrange in a hexagonal manner and will be therefore referred from now on as the 'hexagonal structure'. In comparison with the protrusions observed in the other two island species 'B' and

'C', the protrusions in the hexagonal structure appeared generally brighter.

The second island type, indicated by 'B', appeared to be built up by 'bright rows' separated by one 'dark row' in alternating manner. In order to visualize the concept one of these bright rows is marked with an orange double-arrow in figure 4.2-2, the green double-arrow indicates one of the adjacent dark rows. From figure 4.2-2 it can be derived that a bright row appears bright in the micrograph due to a close ordering of point like protrusions, while a dark row defines the gap to the next bright row. This second island type will be referred from now on as the 'herringbone structure'.

The third recorded island species is defined by its peculiar 'cross structure'. As a reoccurring feature crosses were found which are composed individually by four protrusions, as indicated by the orange circle in figure 4.2-2. These crosses are arranged in a square fashion.

Concerning the distribution of the island types on the surface in general it should be noted that the hexagonal structure was the least occurring, while islands of the herringbone and the cross type were recorded in similar quantity. In addition the size of the herringbone structured islands appeared to be similar to the size of the cross ordered islands. None of the two structures prevailed at medium coverage.

Between the islands stripy features, similar to the ones recorded at low coverage, were found. Furthermore the boundaries of the islands were quite unstable which is related to the observation of attaching and detaching molecules. This will be discussed in more detail for the hexagonal structure later on in this section.

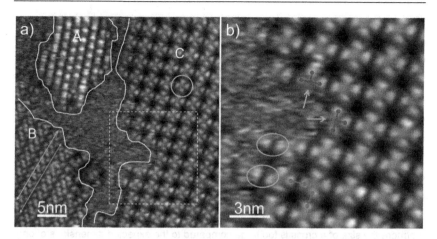

Figure 4.2-2 a) close up of the three different island species. The orange arrow in the hexagonal structured island (A) highlights one of the typical observed bright protrusions. In the herringbone structured island (B) the orange arrow indicates a bright row while the green arrow indicates a dark row. The orange circle in the cross structure (C) illustrates one cross consisting of four protrusions. The dashed white line marks the area examined in **b)**. At the boundary of the cross structured island incomplete crosses consisting of only three protrusions were observed, illustrated by the dark orange drawings. The spots where one protrusion is missing are marked exemplary by the orange arrows. The orange circles indicate that island terminating groups of only two protrusions were found. It is proposed that two protrusions resemble one molecule as illustrated by the dark orange dumbbell drawings. (U_{bias} = +1.21 V, I_{set} = 28.8 pA)

The stripy features between the islands are again the result of 2D gas phase similar as described lower coverage. No individual molecules could be imaged on the surface at RT due to weak molecule-substrate interactions supporting the assumption of a 2D-phase of fast diffusing molecules.[43, 51] Consequently the formation of two dimensional islands is attributed to lateral stabilization of the Ni-TPBP molecules by molecule-molecule interactions. Thus a coexistence of condensed phase and 2D gas phase is apparent at the given coverage.[51]

In order to further analyze the islands and the adsorption behavior of Ni-TPBP the appearance of an individual Ni-TPBP molecule in the scanning tunneling micrographs should be determined. Therefore the arrangement of the protrusions within the islands was analyzed closer since the islands are composed by Ni-TPBP molecules.

On the first glance one might assume that one of the crosses in the cross structure resembles one molecule somewhat similar to the observations reported for Fe-TBP on Cu(111) and Au(111).[54-55] However, at closer inspection of the boundary of the cross island it was possible to find incomplete crosses. Figure 4.2-2 b) presents a cutout of the area marked with the white square in figure 4.2-2 a). The zoom-in of the boundary visualizes that crosses with only 3 protrusions existed indicated by the red drawings. Another observation is indicated by the orange circles. At these spots the cross structure was terminated by a combination of two protrusions. Concerning the molecular appearance it is very unlikely that incomplete molecules are present. Furthermore the size of a cross is too large compared to the expected dimensions of one Ni-TPBP molecule. Thus the assumption of one cross being one molecule can be ruled out. This observation also refutes further possible assumptions of the molecular appearance where the molecule might be composed by a combination of the four protrusions making up one cross.

Since the molecular appearance as one cross can be ruled out it is proposed that the parts of the cross structure, which connect one cross to the next one, resemble one molecule. Combined with the observation of two protrusions terminating the cross structure this bears the idea of two bridging protrusions for the molecular appearance. This approach is in good agreement with the observations made at low coverage. The relevant bridging protrusions are exemplary marked by the dark orange dumbbell shaped drawings in figure 4.2-2 b).

To check this assumption the hexagonal structure was examined more thoroughly. Figure 4.2-3 a) presents scanning tunneling micrograph of a hexagonal structured island. As already stated before the island was composed by bright protrusions that arrange in a hexagonal manner. However, it was also possible to observe numerous features that do not fit to the alignment of the previously observed bright protrusions. In figure 4.2-3 b) a zoomed in part of the center of the island is presented. The orange circle indicates that in this area the hexagonal arrangement of bright protrusions was disrupted by dark protrusions. The adjacent protrusions of the dark protrusion also appeared smaller and less bright than the other protrusions composing the hexagonal structure.

In figure 4.2-3 c)-f) a series of consecutively recorded frames is depicted. The white square in figure 4.2-3 a) marks the area investigated in the consecutive frames while the orange circle indicates the area of interest at the boundary. The red and orange arrows highlight the regions where single Ni-TPBP molecules attached/detached at the island boundary corresponding to 2D gas phase condensation.[51] Interestingly, the attaching/detaching features were always in form of two protrusions. Therefore it is stated that two protrusions are attributed to one Ni-TPBP molecule.

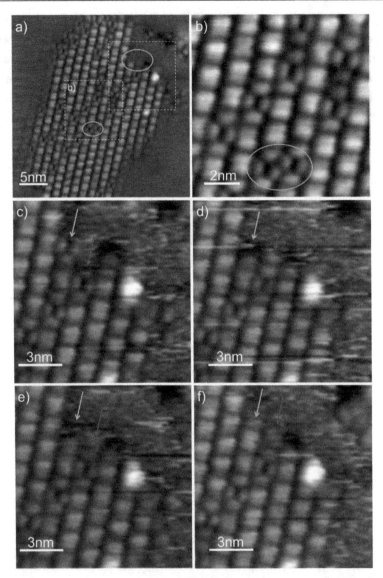

Figure 4.2-3 a) scanning tunneling micrograph presenting a hexagonal structured island. Dark protrusions were observed within the island as well as at the boundary, tagged exemplary by the orange circles. **b)** presents a cut out located at the center of the molecule as indicated by the dashed rectangle. The top right rectangle marks the area investigated in **c)-f)**. The orange and red arrows in the four consecutive frames mark the points of interest for molecule attachment/detachment. (U_{bias} = +1.54 V, I_{set} = 28.2 pA)

4.2.1. Two protrusion model for the three different island types

This section checks for consistency of the two protrusions per molecule model in the three supramolecular arrangements reported for medium coverage. The top row of figure 4.2.1-1 presents close up images of the three different arrangements namely hexagonal, herringbone and cross structure. In the bottom row the same images are presented however with overlaying simplified models to represent the two protrusion model. Since two protrusions are assumed to represent one molecule the alignment of Ni-TPBP molecules in the three different structures can be explained. Note that the orange dumbbell shaped drawings in figure 4.2.1-1 c)-f), used to indicate the molecules, are scaled according to the distance of the respective two protrusions.

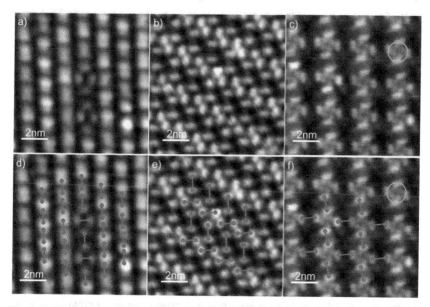

Figure 4.2.1-1 a)-c) present close ups of the hexagonal, herringbone and cross structure respectively. In c) the circle highlights the four small protrusions that were observed in the center of one cross. (a) U_{bias} = +1.54 V, I_{set} = 28.2 pA; b) U_{bias} = +1.65 V, I_{set} = 28.5 pA; c) U_{bias} = +1.20 V, I_{set} = 28.0 pA) In **d)-f)** the dark orange dumbbell drawings mark the positioning of Ni-TPBP molecules in the different structures assuming the proposed two protrusion model.

Figure 4.2.1-1 a) presents a close up of the center of a hexagonal structured island. From section 4.2 it could be derived that two protrusions are assigned to one molecule. In figure 4.2.1-1 d) the corresponding dumbbells are overlayed. The vast majority of the dumbbells are aligned in a hexagonal lattice. In the middle part of the scanning tunneling micrograph two 90° rotated molecules were observed. It should be noted that the models match very well with the supramolecular arrangement indicating that the assumed model is reasonable. The distance between two protrusions of one molecule is roughly 1nm and thus in the range of the expected dimension of one molecule. More information in this regard will be given in the next section. Between the two rotated molecules one single molecule that aligns along the predominant direction was found. In contrast to the other aligned protrusions the protrusions of this molecule were somewhat less bright and not as large. Above the upper 90° turned molecule a molecule was found that contributed to one of the less bright protrusions but also to one of the brighter, larger protrusions. From this the main appearance of the hexagonal structure can be derived. The dark spot between the bright protrusions of one row resembles the center of one Ni-TPBP molecule (see figure 4.2.1-1 d)). Consequently, the bright protrusions are caused by a combination of two fused protrusions of two separate molecules. Therefore the vast majority of the molecules in the hexagonal structured island orientates in a way that the axis defined by the connection of the two protrusions, attributed to one molecule, aligns with the predominant direction of the island. Thus parallel rows result in which the molecules are orientated in the same direction. Further investigations for the hexagonal structured islands will be explained and discussed in section 4.2.5.

The second observed supramolecular arrangement was the so called herringbone structure which is depicted in figure 4.2.1-1 b). As stated before it is defined by 'dark' and 'bright' rows that resemble the predominant orientations of the island. The close up illustrates that the bright rows were due to a close grouping of protrusions while the dark rows were more or less the spacing in between adjacent bright rows. Figure 4.2.1-1 e) introduces the orientation of molecules in the herringbone structure consistently. Considering the two protrusion per molecule model two different proposed orientations of Ni-TPBP molecules are found which are azimuthally rotated by 90° in respect to each other. The molecules located in a dark row face the center of the molecules located in the adjacent bright rows while the molecules of the bright rows face the periphery of dark row molecules. Thus the observed appearance of the bright rows is due to a close grouping of the protrusions of the molecules while the dark rows are caused by the dark centers of the center of the molecule models. Through the close packing of molecules a herringbone orientation results similar to reports of Co-TTBPP on Ag(111).[12] The proposed model orientation is the only reasonable one because other imaginable orientations would not fit for the expected dimensions of the molecule. The herringbone structure will be discussed in more detail in section 4.2.6.

Finally the molecule orientation in the cross structure is discussed briefly. Figure 4.2.1-1 c) presents a close up of a center of a cross structure, while figure 4.2.1-1 f) presents the overlaying models. Characteristically for the cross structure four bright protrusions per cross were observed. However, as stated before, one molecule resembles the bridge between two crosses as illustrated in figure 4.2.1-1 f). The centers of the molecules are located in the dark area between two bright protrusions analog to the arrangement in the hexagonal and herringbone structure. Furthermore the center of the molecules align along one of the two predominant directions of the cross structure which are rotated by 90° in respect to each other. The bridging molecules are rotated counter clockwise in respect to the predominant directions of the observed island. Interestingly, in the scanning tunneling micrograph further four small protrusions in the center of one cross were observed, indicated by the orange circle in figure 4.2.1-1 c) and f). This appearance cannot be explained with the two protrusion model but by an additional molecule, featuring a different intramolecular conformation, located in the center of a cross. This will be discussed in more detail along with the cross structure in section 4.2.7.

4.2.2. Intramolecular conformation for the 'two protrusions' appearance

In the following section the intramolecular conformation of Ni-TPBP determining the observed 'two protrusions' molecular appearance will be discussed. For porphyrins tilt and twist angle are commonly used to describe the intramolecular conformation. Figure 4.2.2-1 presents which tilt and twist angles are the important parameters to understand the conformation of Ni-TPBP. For demonstration only one of the four tetraphenyl legs and only the tilt angle of one of the four benzopyrrole groups is depicted. However the approach can be also applied to any substituents.

Generally the tilt angle describes the inclination of a group in respect to a certain plane. In figure 4.2.2-1 a) the tilt angle of one of the benzopyrrole groups is illustrated. In this case the benzopyrrole group is rotated out of the macrocycle plane around the axis which is defined by the virtual connection between alpha positioned atoms of the respective pyrrole group (see dashed blue line in figure 4.2.2-1 a)). Figure 4.2.2-1 b) presents an example for a tilting of the benzopyrrole group out of the macrocycle by $\alpha_{bpy} = 30°$. The twist angle describes a rotation of a substituent. In the case of Ni-TPBP the rotation of a phenyl ring around the axis defined by the C-C connection to the macrocycle is of interest (see dashed red line in figure 4.2.2-1 a)). By definition a twist angle of 0° for the phenyl ring describes a flat conformation of the phenyl ring in plane with the macrocycle (disregarding any tilt angle of the respective phenyl leg). In figure 4.2.2-1 b) an exemplary twist angle of $\alpha_{ph} = 60°$ is depicted.

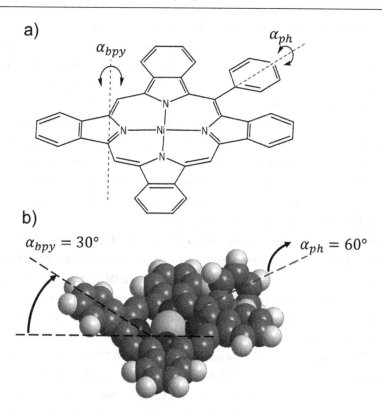

Figure 4.2.2-1 a) model illustrating the tilt angle of the benzopyrrole group (α_{bpy}) and the twist angle of the phenyl group (α_{ph}) in Ni-TPBP. The model only shows one phenyl leg for better visibility of the rotation axes and angles. b) space filled model of a) after rotation of α_{bpy} = 30°and α_{ph} = 60°.

Generally for TPPs a saddle shaped conformation of the molecules has been report-ed similar to the one explained for 2HTPP in section 2.4.2.[41-43] Characteristic for the saddle shape is the inclination of two adverse pyrrole groups towards the surface while the remaining two pyrrole groups are bent upwards. The magnitude of tilt and twist angle depends on the respective TPP and the corresponding substrate. Since the constitution of Ni-TPBP is very close to those of M-TPPs a saddle shape conformation for Ni-TPBP on Cu(111) is reasonable. Figure 4.2.2-2 b) illustrates an exemplary saddle shape conformation for Ni-TPBP. The tilt angle for the benzopyrrole groups and the twist angle of the phenyl legs was adapted from the values reported for Cu-TPP on Cu(111).[42] The important differences in Ni-TPBP compared to Cu-TPP, apart from the metal center, are the fused benzene groups at the beta positions of the pyrrole rings. Under the assumption that the appearance in STM follows the topography of the molecule the upwards bent benzopyrrole groups are most likely to dominate the appearance of Ni-TPBP in STM. Therefore in the thesis at hand the respective groups will be highlighted by yellow color as illustrated in figure 4.2.2-2 b). Consequently the two protrusions, which were assumed to accord to one molecule in section 4.2, can be attributed to two opposing benzopyrrole groups. Figure 4.2.2-2 a) presents the aforementioned typical molecular appearance of Ni-TPBP featuring two bright protrusions and a dark center.

At this point the appearance of the metal center of Ni-TPBP on Cu(111) in STM shall be described briefly. Bias dependent appearance of the metal center has been reported by Buchner et al. for Co-TPP on Ag(111) in STM.[45] For a given voltage Co-TPP shows a central protrusion while metalloporphyrins such as Fe-TPP show the typical saddle shape contrast. This behavior was attributed to the electronic state resulting from interactions of the half-filled d_{z^2} orbital of the cobalt center and the surface.[45] However, for nickel center porphyrins, such as Ni-TPBP or Ni-TPP, no such behavior is expected since the central nickel atom has a filled d_{z^2} orbital.[56-57] Therefore the metal ion valence configuration has a strong influence on the observed tunneling images.[56, 58] Considering the molecular appearance of Ni-TPBP on Cu(111) in STM, as presented in figure 4.2.2-2 a), the metal center appeared as a dark protrusion and showed no pronounced bias dependence.

a)   b)

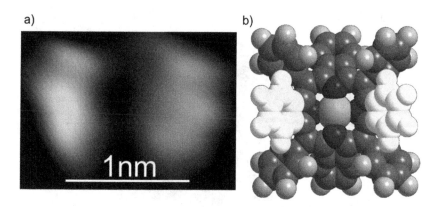

Figure 4.2.2-2 a) Typical two protrusion contrast for one Ni-TPBP molecule on Cu(111). (U_{bias} = +1.31 V, I_{set} = -29.5 pA) **b)** Exemplary saddle shape model proposed for Ni-TPBP. Note that the parameters for tilt angle of the benzopyrrole groups and the twist angle of the phenyl legs were adapted from Cu-TPP on Cu(111).[42] The benzopyrrole groups that are proposed to dominate the molecular appearance are colored yellow.

Since the values of the tilt and twist angles used for the Ni-TPBP model in figure 4.2.2-2 b) were simply adapted from Cu-TPP a more specific model for the intramolecular conformation of Ni-TPBP still has to be determined.

One way of determining the tilt and twist angles and thus the intramolecular conformation of a molecule is measuring certain distances in the recorded STM image and using a model of the molecule as well as basic trigonometry. The choice of measured distances depends on the molecular appearance in STM. As stated before the benzene rings of two adverse benzopyrrole groups are proposed to dominate the molecular appearance. However, it is important to note that only the tilt angle of the benzopyrrole groups can be determined since the contrast observed in STM gives no information concerning the phenyl legs of Ni-TPBP. In figure 4.2.2-3 the procedure of identifying the tilt angle of the benzopyrrole groups is depicted. For better visibility of the tilted benzopyrrole groups the phenyl legs are left out in the model.

a)

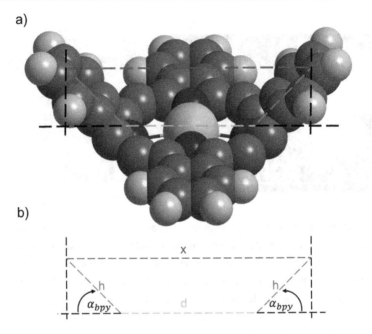

b)

Figure 4.2.2-3 a) space filling model of Ni-TPBP where two adverse benzopyrrole groups are bent upwards. Note that the phenyl groups are left out for clarity. The colored lines indicate the distances relevant for calculation of the tilt angle α_{bpy}. **b)** presents a sketch of the distances shown in a).

The marked distances from the space filling model of Ni-TPBP in figure 4.2.2-3 a) are shown again in figure 4.2.2-3 b) to illustrate the applied deformation and the notation. By 'd' the distance between the rotation axes of adverse benzopyrrole groups is described. This distance is a constant value and is read out from the model using the program chem3D. The other set value 'h' is read out likewise. It constitutes the distance from the former mentioned rotation axes of a benzopyrrole group to the top part of the benzene ring of the respective benzopyrrole group. The value of 'x' is defined by measuring the distance of the two protrusions recorded in the scanning tunneling micrographs. By applying triangulation the tilt angle α_{bpy} is calculated:

$$\cos(\alpha_{bpy}) = \frac{(x - d)}{2 \cdot h}$$

Measuring the distances of the two protrusions for several molecules in different scanning tunneling micrographs results in a mean value of a tilt angle for the benzopyrrole groups of $\alpha_{bpy} \approx 45°$. Figure 4.2.2-4 a) presents a model incorporating the calculated value. Since the twist angle of the phenyl legs cannot be determined an angle of $\alpha_{ph} \approx 60°$ is assumed similar to the results of Cu-TPP on Cu(111).[42] A sideview of the model can be seen in figure 4.2.2-4 b). The rotation of the phenyl legs decreases the destabilizing effect of steric repulsion in respect to the adjacent benzopyrrole groups.

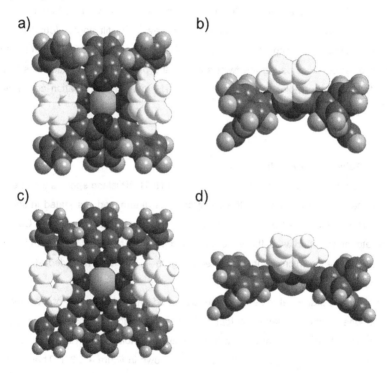

Figure 4.2.2-4 a) proposed saddle shape model for the intramolecular conformation of Ni-TPBP on Cu(111) with a tilt angle of 45° for two opposite benzopyrrole groups. The twist angle of the phenyl legs is 60° adapted from reported twist angles of TPPs. **b)** sideview of the model in a). **c)** DFT energy minimized gas phase model for the intramolecular conformation of Ni-TPBP on Cu(111). The saddle shape is defined by a tilt angle of $\alpha_{bpy} \approx 40\pm5°$ for the benzopyrrole groups and a twist angle of $\alpha_{ph} \approx 50\pm5°$ for the phenyl legs. Calculations were performed by the group of Wolfgang Hieringer. **d)** sideview of the model in c).

Furthermore figure 4.2.2-4 c) contains the results of a DFT calculated gas phase model for Ni-TPBP. Calculations were performed by the group of Wolfgang Hieringer. The model resembles the energetic global minimum. Clearly the intramolecular conformation resembles a saddle shape supporting the assumption made beforehand. The sideview in figure 4.2.2-4 d) indicates that two adverse benzopyrrole groups are bent upwards by $\alpha_{bpy} \approx 40\pm5°$ while the remaining two are bent downwards equally. The twist angles of the phenyl groups in the gas phase model constitute $\alpha_{ph} \approx 50\pm5°$. It must be emphasized that the energy optimized model does not incorporate the influence of the surface and intermolecular interactions. Thus deviations from the model calculated by triangulation are expected. However, the tilt angles found for the model calculated by triangulation and the DFT model are in good agreement in the margin of error. Considering the twist angle of the phenyl legs deviation has to be expected since the molecular appearance in STM gives no information concerning the phenyl groups.

4.2.3. Substrate orientation

In order to further characterize the three different Ni-TPBP island species the orientation of the respective islands to the substrate was determined. As stated in section 2.4.2 2HTPP is a suitable marker for the high symmetry directions of the Cu(111) substrate at RT, and was therefore co-adsorbed with Ni-TPBP on Cu(111). The scanning tunneling micrograph presented in figure 4.2.3-1 was recorded at medium coverage of Ni-TPBP.

On the surface individual molecules were observed. In the upper part of the image a herringbone structured island was found. From investigation of consecutive images it was possible to track the movement of the individual molecules on the surface. The diffusing directions are illustrated by the white arrows in figure 4.2.3-1. They resemble multiples of 120° indicating the characteristic substrate orientation reported for 2HTPP on Cu(111).[13, 46] Therefore the diffusing directions of the individual molecules in figure 4.2.3-1 correlate to the high symmetry directions of the substrate marked by the white arrows. Interestingly, the two protrusions of the molecules of the dark rows are aligned accordingly to one of the high symmetry directions. Again orange colored dumbbell models are used in figure 4.2.3-1 b) to illustrate the molecules of the herringbone structure and their orientation to the substrate.

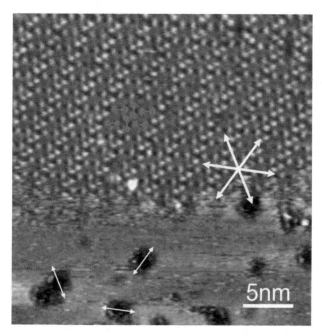

Figure 4.2.3-1 Scanning tunneling micrograph presenting a herringbone structured Ni-TPBP island with co-adsorbed 2HTPP on Cu(111). (U_{bias} = -0.453 V, I_{set} = 29.9 pA) The arrows mark the high symmetry axes of the substrate determined by the individual observed 2HTPP molecules. From the dumbbell illustration for the Ni-TPBP molecules in the island it can be derived that the two protrusions of dark row molecules align along one of the substrate directions.

As the substrate orientation of the herringbone structured islands has been determined the knowledge can be transferred to the hexagonal and the cross structured islands. Therefore the micrograph in which all three different species were recorded at the same time is examined in figure 4.2.3-2. From the orientation of the molecules in the dark rows of the herringbone structure the high symmetry directions of the substrate can be determined. For the cross structure one of the lattice vectors coincides with one of the high symmetry substrate directions. In case of the hexagonal structure the molecules align with some deviation with the high symmetry Cu(111) directions. The different orientations of the three island species will be discussed in more detail in section 4.2.5, 4.2.6 and 4.2.7 respectively.

Figure 4.2.3-2 Three different island species recorded in one scanning tunneling micrograph. (U_{bias} = +1.21 V; I_{set} = -28.8 pA) The dumbbell drawings indicate the arrangement of molecules in the respective islands. The two protrusions of the dark row molecules of the herringbone structure were used to determine the substrate orientation. In the cross structure one of the lattice vectors coincides with one substrate direction while in the hexagonal island the predominant direction does.

4.2.4. Dynamic behavior of Ni-TPBP on Cu(111)

In this section the dynamic behavior of Ni-TPBP on Cu(111) will be briefly discussed. Within the three different reported supramolecular structures no particular dynamic behavior of the molecules was observed. The molecules within the island were stable and exhibited no obvious rotational movement or switching as observed for example for 2HTTBPP on Cu(111).[59]

Furthermore compared to the dynamic behavior of individual 2HTPP molecules at RT, which diffuse along the three high symmetry axes of the substrate, the diffusion of single Ni-TPBP molecules on the surface was not possible to track at RT. Generally, the diffusion rate obeys Arrhenius law.[46, 60] For 2HTPP the strong interaction between molecules and surface lead to the characteristic movement along the main crystallographic directions similar to a 'train on rails'.[41] However, for Ni-TPBP on Cu(111) no such behavior was observed indicating that the interaction of molecules with the substrate is probably low compared to 2HTPP. At low coverage adsorbed Ni-TPBP molecules cannot be stabilized by lateral interactions with other molecules leading to the observation of 2D gas phase and step decoration.[52-53] The quickly diffusing molecules in the 2D gas phase were just observable in the images in form of stripes because the diffusion speed of the molecules exceeds the recording speed of the STM. The exact speed of diffusion depends on several factors like molecule-substrate interaction, molecule-molecule interaction, temperature, coverage and density of defects. Increasing the number of Ni-TPBP molecules in the 2D gas phase by depositing further material to a critical number lead to the formation of two dimensional islands that were separated by a phase of 2D gas. Mutual lateral stabilization of the benzoporphyrin molecules accounts for the formation of islands. However, the boundaries of the islands were reported to be instable at medium coverage. As explained in section 4.2 the instability was caused by molecules that detach from the island to 'join' the 2D gas phase, as well as by molecules that attached to the supramolecular islands corresponding to 2D gas phase condensation.[51]

Considering the stability of the islands at medium coverage it was observed that the islands were easier to image in high positive bias. Similar behavior has been reported for Co-TPP on Cu(111).[61] By applying high voltages and low currents a high 'tunneling resistance' and thus a large tip-surface distance results leading to a small interaction between tip and molecules.[48, 62] On the other hand for lower tunneling resistances the tip is closer to the surface and thus transformation of the ordered islands to a disordered state is promoted. Therefore the disordered state resembles highly mobile Ni-TPBP similar to the diffusing molecules of the 2D gas phase.

4.2.5. Hexagonal structure

The observation of three different island species at the same area and RT was characteristic for medium coverage of Ni-TPBP on Cu(111). One type of these structures were the hexagonal structured islands that are defined by a hexagonal arrangement of bright protrusions. In comparison to the protrusions of the other two island species the protrusions building up the hexagonal structured islands appeared brighter. Furthermore hexagonal structured islands were observed rarely and only in form of relative small islands compared to the herringbone and cross structure. Considering stability of the islands the hexagonal islands were reported to constitute instable boundaries attributed to detaching/attaching of molecules as presented in figure 4.2-3. According to the two protrusions model proposed in 4.2.1 the Ni-TPBP molecules in this structure order hexagonally. It was suggested that two molecules of one row contribute to the characteristic bright protrusion contrast. In section 4.2.3 it was identified that the predominant direction of the hexagonal structured island coincides with one of the high symmetry axes. Interestingly, within the island as well as at the boundary molecules were observed that are rotated by 90° in respect to the predominant direction of the island.

As stated in section 4.2.2 the molecular appearance of Ni-TPBP on Cu(111) was proposed to be dominated by the two upwards bent benzopyrrole groups of a saddle shaped intramolecular conformation. In order to interpret the hexagonal structure more thoroughly the orientation of molecules within the island species combined with molecule models incorporating the intramolecular conformation is needed. Therefore high resolution images of the hexagonal structure were recorded and examined.

Figure 4.2.5-1 presents such a high resolution scanning tunneling micrograph of the hexagonal structure island. The image is superimposed with scaled molecular models incorporating the aforementioned saddle shape conformation. For the tilt angle of the benzopyrrole groups 45° were calculated while the twist angle for the phenyl legs of 60° was transferred from results of similar systems. The orientation of the models is according to the discussion in section 4.2.1. From figure 4.2.5-1 it becomes clear that a combination of two upwards bent benzopyrrole groups of two molecules contribute to one bright protrusion. Thus parallel rows of molecules are formed in which the two upwards bent benzopyrrole groups of the individual molecules are orientated along the predominant direction of the island.

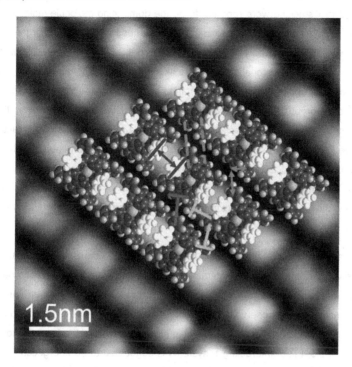

Figure 4.2.5-1 High resolution scanning tunneling micrograph of the hexagonal structure superimposed with molecule models. Possible interactions are exemplary indicated by different colors and drawings: The dark red parallel lines illustrate π-π stacking between benzopyrrole groups. The dark red arrow marks the center to center distance of respective groups. To visualize π-π stacking between phenyl legs of adjacent row molecules the orange parallel lines are given. The orange arrows indicate the center to center distance as well as the center displacement of interacting groups. T-type interaction between phenyl legs and benzopyrrole groups is marked with a red shaped T. On the other hand T-type interaction of phenyl legs is indicated by a green T-shape. (U_{bias} = +1.54 V, I_{set} = 28.2 pA)

Considering molecular interactions within the hexagonal structured islands one molecule has the possibility to interact with its surrounding six molecules. Generally for agglomerated islands of porphyrin molecules and their derivatives, T-type interactions and π-π stacking were reported to be the dominating molecule-molecule interactions.[11, 43, 63] The existence and importance of these interactions in organic molecules have been discussed extensively.[64] In figure 4.2.5-1 the proposed interactions for Ni-TPBP in the hexagonal structure are illustrated. Basically four different kinds of interaction possibilities are proposed. It has to be noted that the following discussions of interacting group positions as well as measured distances are limited by the assumed molecule models. Especially for the influence of phenyl legs on interactions it has to be kept in mind that their conformation could not be determined from the STM appearance and had to be adapted from reports of a similar system. However, the adapted angles are in good agreement with the results of the DFT calculated model. Therefore the trends and approximations describe the system qualitatively correct and thus can be discussed and compared.

One potential interaction of interest is π-π stacking of two benzene rings of upwards bent benzopyrrole groups. The interaction is illustrated exemplary with two dark red colored parallel lines in figure 4.2.5-1. It seems reasonable that such an interaction exists considering the observation of bright protrusions that were attributed to a combination of two benzoyprrol groups. For π-π stacking an approximate center to center distance of 3.5-4 Å of interacting π-systems is suggested to be energetically favorable.[65-66] The center to center distance regards the separation distance of interacting approximately parallel molecular planes. In case of the hexagonal structured island the mean center to center distance for upwards bent benzopyrrole rings, indicated by the red arrow in figure 4.2.5-1, yields 6.7±0.39 Å. Therefore this kind of interaction can be immediately ruled out.

Another π-π stacking interaction might occur between phenyl legs of adjacent row molecules as indicated by the parallel orange dashed lines in figure 4.2.5-1. In this case the center to center distance and more importantly the displacement of the centers from the 'face to face' orientation of the phenyl legs have to be considered.[65-66] The orange arrows in figure 4.2.5-1 indicate the distances of interest. The center displacement distance describes the distance of the center of interacting groups within one molecular plane. Sherill et al. state that an optimal distance for the center displacement equals 1.6 Å.[66] Calculating the mean center to center distance for interacting groups yields 4.3±0.36 Å while the mean center displacement constitutes 5.3±0.57 Å. Considering this rather large distance it is not reasonable to propose π-π stacking for the phenyl groups.

The other type of attractive interactions in agglomerated porphyrin islands are T-type interactions which are defined by an 'edge to face' positioning of involved groups.[43] In the case of the hexagonal structure one possible T-type interaction is induced over the para-positioned carbon of a phenyl leg to the downwards bent benzene ring of the benzopyrrole group of an adjacent row molecule. The interaction is illustrated by the red colored T-shape in figure 4.2.5-1. Petsko et al. stated that T-type interacting aromatic groups have a preferred separation distance from center to center of between 4.5 to 7.0 Å.[67] The range of preferred separation distances is in good agreement with calculations of Sherrill et al.[66] In the case of the proposed interaction within the hexagonal structured island a mean distance of 6.9±0.86 Å was determined. This value is at the upper limit of the reported preferred separation distances but is still reasonable considering flexibility of groups and the accuracy of the given models. Moreover, for the dihedral angle of the 'edge to face' positioning Petsko et al. found a range of preferred angles between 50° to 90°. The dihedral angle for the proposed T-type interacting groups varies because some of the molecules in hexagonal structure deviate from the hexagonal lattice. It should be noted that the groups are not fixed but somewhat flexible and the mean values were just calculated from the molecule model that is based upon the calculated tilt angle of the benzopyrrole group as well as a twist angle of the phenyl legs that was transferred from similar systems. Therefore it is proposed that the interaction is improved by reorientation of interacting groups. Considering these circumstances the discussed T-type interaction is reasonable and would contribute to the stability of the hexagonal islands.

In addition T-type interaction between one of the meta-positioned carbons of a phenyl leg to the center of the phenyl leg of a neighboring molecule is proposable. The corresponding interactions are illustrated exemplary for one case with green colored T-shapes in figure 4.2.5-1. The mean center to center distance for such an interaction constitutes 6.1±0.44 Å which is in excellent agreement with the values reported in literature.[65-68] Overall the importance of this interaction is hard to judge due to the limitations of the model and the irregular distances of molecules in the hexagonal structure.

All in all the most reasonable contributions to the molecule-molecule interactions in the hexagonal structured islands are in form of T-type interactions while π-π stacking interactions were ruled out because of deviating geometry and too large center to center distances. Figure 4.2.5-2 presents an overview of the interactions considering just the aforementioned T-type interactions. The colored lines symbolize the connections between center to center of interacting groups. Due to the red colored T-type interactions, induced by molecules of adjacent rows, and the green colored T-type interactions 'zigzag' shapes for the correlation of interactions result. Obviously the zigzag shapes are distorted and irregular which might indicate that the molecules in the hexagonal structure are not stable, but try to arrange in a more favorable way to improve interactions. This is in good agreement with the observation of turned molecules within the island which possibly resembles a transition state of the island towards the herringbone or the cross structure. Another indication for the low stability of the hexagonal structured islands is that they were only rarely observed while herringbone and cross were regularly found in form of larger islands. It should be noted that the proposed interactions are based on a model that was calculated from the distances of the characteristic two protrusion appearance. However, it explains the observations qualitatively and is therefore a good approximation.

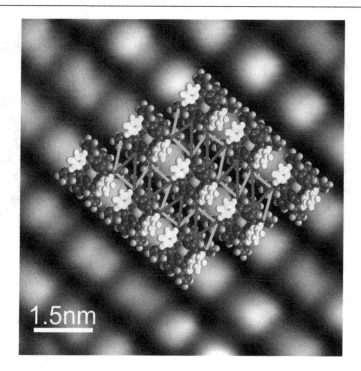

Figure 4.2.5-2 High resolution scanning tunneling micrograph of the hexagonal structure superimposed with molecule models. The red lines symbolize the T-type interaction between phenyl and benzopyrrole groups, the green lines T-type interaction between phenyl legs. (U_{bias} = +1.54 V, I_{set} = 28.2 pA)

As a next step of characterizing the hexagonal structured islands the unit cell and the corresponding lattice parameters of the structure were determined. Figure 4.2.5-3 a) presents an average frame recorded in the center of a hexagonal structured island. The orange lines indicate the dimensions used to determine the lattice parameters of the hexagonal unit cell. An exemplary unit cell is indicated by the black rectangle in figure 4.2.5-3 a). The hexagonal unit cell is defined by the unit cell vectors a = 1.47±0.09 nm and b = 1.63±0.07 nm as well as an angle α = 116±5°. Considering the proposed two protrusion model the unit cell contains one molecule yielding a molecular density of 0.48 mol./nm².

Figure 4.2.5-3 a) scanning tunneling micrograph of the hexagonal structure. The unit cell is defined by $a = 1.47\pm0.09$ nm, $b = 1.63\pm0.07$ nm and $\alpha = 116\pm5°$. ($U_{bias} = +1.65$ V, $I_{set} = 28.5$ pA) **b)** high resolution image of the hexagonal structure with overlayed molecule models. The arrows mark the high symmetry substrate directions. The predominant direction of the hexagonal structure coincides with one substrate direction. ($U_{bias} = +1.54$ V, $I_{set} = 28.2$ pA)

Another point of interest is the orientation of the hexagonal structured islands in respect to the substrate. In section 4.2.3 it was found that the predominant direction of the hexagonal island coincides with one of the high symmetry axes of the substrate. Thus three different possible orientations arise for the hexagonal structured islands. However, since this type of island is only rarely observed, it was not possible to record the different orientations within one image. Regarding the ordering of molecules according to the substrate, figure 4.2.5-3 b) presents a high resolution image of a hexagonal structured island with overlaying molecular models. From the predominant direction of the island the high symmetry directions of the substrate, marked by the black arrows, can be determined. Considering the predominant direction the molecules align in good agreement with the high symmetry axes but for the remaining two directions derivations from the substrate orientations are apparent. This is caused by the not perfect hexagonal ordering of the island induced probably by the overall stability of the structure. For the molecules within one row of the predominant direction it is assumed that some attractive interaction between molecules and substrate atoms exist. However, it was not possible to determine the cause of this interaction from the recorded images.

4.2.6. Herringbone structure

The second supramolecular arrangement that will be discussed in more detail is the herringbone structure. In general the herringbone structure was more frequently observed than the hexagonal type and in larger islands indicating an energetically favorable situation. Characteristic for this arrangement were observations of alternating bright and dark rows within the islands which resemble the predominant directions of the respective island. As explained in section 4.2.1 for herringbone structured islands Ni-TPBP molecules are orientated in a way that molecules in a dark row face the center of respective bright row molecules while bright row molecules face the periphery of respective dark row molecules (see figure 4.2.1-1 e). In respect to the dark row molecules the bright row molecules are rotated by 90°. Thus a close packing of molecules occurs which resembles a herringbone structure. In section 4.2.3 it was determined that the two protrusions of dark row molecules align along one of the high symmetry axes of the substrate.

High resolution images of the supramolecular arrangement with overlayed molecule models are presented in figure 4.2.6-1. The orientation of molecules is according to section 4.2.1 while the molecule models correspond to the saddle shape model from section 4.2.2 calculated from the distance of two protrusions. Clearly the bright rows appear as bright due to close groupings of upwards bent benzopyrrole groups of neighboring molecules while the 90° turned dark row molecules resemble spacers to the next bright row. Interestingly, the orientation of molecules in the herringbone structure is somewhat related to the hexagonal structure. The structural difference is that the molecules of adjacent rows in the herringbone structure are turned by 90° to each other. Therefore a 'gear teeth' like arrangement of the benzopyrrole groups results. Besides the order of molecules in the herringbone structure is more regular which promotes close packing and thus shorter and more regular distances between molecules compared to the hexagonal structure.

In order to understand the observations for the hexagonal structured islands, the interactions of molecules within the structure have to be discussed. Therefore possible

T-type and π-π stacking interactions have to be examined.[11, 43, 63] Considering the positions of molecules in the herringbone structure, one molecule has the possibility to interact with six neighboring molecules similar to the hexagonal structure. Examining the interactions between the molecules leads to 5 different possibilities. Since the herringbone structure is quite similar to the hexagonal structure, similarities for the molecule-molecule interactions are expected.

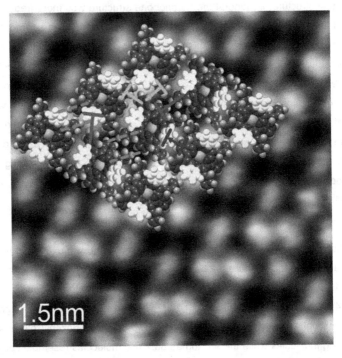

Figure 4.2.6-1 High resolution scanning tunneling micrograph of the herringbone structure superimposed with molecule models. Possible interactions are exemplary marked by different colors and shapes: The dark red parallel lines symbolize the π-π stacking between benzopyrrole groups. The center to center distance is marked by the dark red arrow. The π-π stacking between phenyl legs of molecules within either bright or dark row are illustrated by the orange parallel lines. For marking the π-π stacking of phenyl legs of adjacent row molecules purple parallel lines are given. The T-type interactions between phenyl legs and benzopyrrole groups of neighboring molecules are presented in form of red T shapes. Whereas the T-type interactions between phenyl legs of neighboring molecules are illustrated by green T shapes. (U_{bias} = +1.65 V, I_{set} = 28.5 pA)

The first possible interaction that should be discussed is π-π stacking between the benzene ring of an upwards bent benzopyrrole group of a dark row molecule with the benzene ring of a downwards bent benzopyrrole group of the adjacent bright row molecule. To visualize the concept the interaction is marked exemplary with two dark red parallel lines in figure 4.2.6-1. Since the upwards bent benzopyrrole groups of dark row molecules face the center of adjacent bright row molecules there is no center displacement of the interacting benzene rings. Also the geometry of the interacting groups suggests good interaction because they are positioned 'face to face'. However, the mean center to center distance yields 4.9±0.16 Å. As stated before the favorable center to center distance is in the regime of 3.5-4 Å.[65-66] Therefore it is unlikely that the proposed π-π stacking is reasonable.

Another stacking interaction between π-systems could occur between either two phenyl legs of neighboring dark row molecules or two phenyl legs of neighboring bright row molecules. The proposed interactions are illustrated by the dashed orange lines in figure 4.2.6-1 for both possibilities. Considering the geometry and distances of involved groups the two possibilities are similar and can be discussed as one. The characteristics of the interaction are quite similar to the π-π stacking that was proposed for phenyl legs of adjacent rows in the hexagonal structure. Again steric repulsion of the phenyl legs and neighboring benzopyrrole groups hinders and limits altering of the twist angle which is necessary to achieve 'face to face' orientation and thus improved interaction of the phenyl legs. While it might be possible that approximate 'face to face' geometry is achieved the mean center to center distance for the groups of interest constitutes 4.5±0.34 Å which is too large for favorable interaction.[65-66] In addition the mean center displacement yields 3.1±0.16 Å which further resembles an unfavorable positioning of the phenyl legs for π-π stacking interactions.[66] Regarding the different aspects of the proposed interaction, π-π stacking of the phenyl legs is not reasonable.

The last possible π-π stacking interaction that shall be discussed is the interaction of a phenyl leg of a dark row molecule with a phenyl leg of a bright row molecule. The purple dashed lines in figure 4.2.6-1 indicate the interactions. Because of the 90° rotated orientations of dark and bright rows molecules the phenyl legs for the proposed interaction are positioned in the favorable 'face to face' orientation. However, the interactions has to be ruled out because the mean center to center distance of the phenyl legs is 5.4±0.14 Å and the mean center displacement is 3.2±0.22 Å. Thus the distances between the π-systems are clearly too large for favorable interactions.

As for the hexagonal structure the different imaginable possibilities for π-π stacking between molecules in the herringbone structure are not reasonable. To complete the investigation possible T-type interactions have to be discussed.

The first T-type interaction to be discussed is between the para-positioned carbon of one phenyl leg of either one bright or one dark row molecule to the center of a benzene ring of a bent down benzopyrrole group of a dark row molecule. The interaction is illustrated by a red colored T-shape in figure 4.2.6-1. As it is indicated in the figure interaction of phenyl legs to the center of a benzene ring of an upwards bent benzopyrrole ring of a bright row molecule is also possible. Since the geometries and distances of involved groups are comparable, the two different possibilities for this kind of interaction will be discussed as one. The mean center to center distance yields 6.2±0.25 Å which is in line with the regime of 4.5-7.0 Å stated by Petsko et al. as well as with the calculations of Sherrill et al.[66-67] Considering the orientation of interacting groups it is very likely that due to the flexibility of groups the favorable regime defined by a dihedral angle of 50-90° is achieved.[67] Thus this kind of T-type interaction is plausible and will contribute to the stability of herringbone structured. It should be noted that this kind of interaction is very similar to the T-type interaction between phenyl legs and benzopyrrole rings proposed for the hexagonal structure. Another T-type interaction proposed for the herringbone structure is very similar to the second T-type interaction reported for the hexagonal structure. The proposed interaction takes place between one of the meta-positioned carbons of a phenyl leg of a dark row molecule to the center of the phenyl leg of a neighboring molecule. Green T-shapes in figure 4.2.6-1 mark the interactions exemplary for one phenyl group of a dark row molecule. Interestingly, the mean center to center distance for the involved phenyl legs constitutes 5.6±0.29 Å. This is in excellent agreement with the favorable distances for interacting aromatic systems reported in literature.[66-68] Considering the positioning of the phenyl legs for the proposed interaction, a change of the twist angle of the phenyl legs is necessary to achieve approximate edge to face orientation. However, steric repulsion of phenyl rings and benzopyrrole groups limits such a rotation. If steric repulsion and improvement of interactions can be balanced this interacting is reasonable to contribute to the molecule-molecule interactions.

All in all in the herringbone arrangement, T-type interactions are the most likely ones to contribute to the stability of the structure. Figure 4.2.6-2 presents a summary of the plausible interactions. Interestingly, the proposed T-type interactions are very similar to the ones found for the hexagonal structure. However, in the herringbone structure the interacting groups are positioned in a more favorable way compared to the hexagonal structure. This is represented by shorter distances of the red parts of the 'zigzag' shapes in figure 4.2.6-2. Furthermore the molecules in the herringbone structure are more regularly ordered and thus the interactions are equally distributed between molecules. The more favorable positioning of molecules in the herringbone structure is probably related to the 90° turned orientation of molecules in the dark and bright rows. The 'gear teeth' like orientation of benzopyrrole groups minimizes the steric repulsion between the molecules and at the same time yields a closer distance of T-type interacting groups. Moreover for the herringbone structure a reorientation of interacting groups does not negatively interfere with other interactions which resembles a more stable situation as it was determined for the hexagonal structure. The assumption that the herringbone structure resembles a more favorable structure than the hexagonal structure is also in good agreement with the observations for the adsorption behavior. Within the hexagonal structure defects in form of rotated molecules were reported which might indicate a transition of the molecular order towards the arrangement found in the herringbone structure. In addition the herringbone structure was observed more often in form of larger islands while the hexagonal structure was only rarely observed.

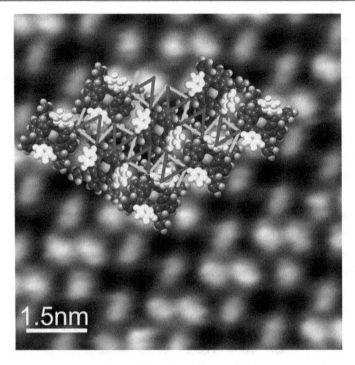

Figure 4.2.6-2 High resolution scanning tunneling micrograph of the herringbone structure superimposed with molecule models. The red lines symbolize the T-type interaction between phenyl and benzopyrrole groups of neighboring molecules, the green lines the T-type interaction between phenyl legs of neighboring molecules. (U_{bias} = +1.65 V, I_{set} = 28.5 pA)

Favorable interaction of the molecules and minimization of steric repulsion due to the orientation of molecules is also represented in the closer packing of the molecules. This is resembled by the molecular density calculated for the unit cell of herringbone structured islands. Figure 4.2.6-3 a) presents a scanning tunneling micrograph of a herringbone structured island. For the herringbone structure the unit cell is also hexagonal. The orange lines illustrate the positions of the height profiles that were used to calculate the unit cell vectors. For the first unit cell vector the mean value constitutes a = 1.55±0.05 nm, the second unit cell vector equals b = 2.81±0.1 nm. The angle defining the hexagonal unit cell amounts α = 115±1°. Since the unit cell contains two molecules a molecular density of 0.52 mol./nm² results. As expected this value is higher than the molecular density of the hexagonal structured islands.

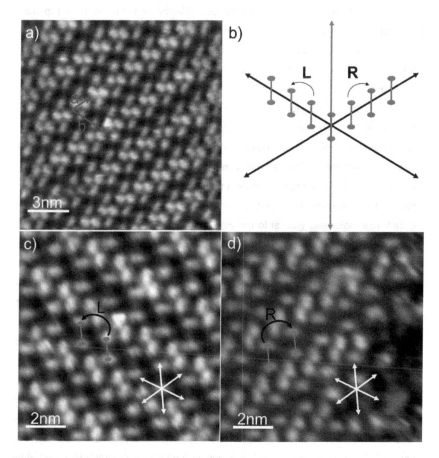

Figure 4.2.6-3 a) scanning tunneling micrograph of the herringbone structure. The unit cell is defined by a = 1.55±0.05 nm, b = 2.81±0.1 nm and α = 115±1°. (U_{bias} = +1.65V, I_{set} = 28.5 pA) **b)** Drawing used to indicate the chirality of the herringbone structure. The green arrow indicates the substrate orientation for which the dark row molecules, indicated by the dumbbell shapes, are aligned. Considering the arrangement of molecules in the herringbone structure two different orientations, namely (L) and (R), are possible for the dark row molecules, as indicated by the black arrows. **c)** example of an (L) herringbone island. (U_{bias} = +1.65V, I_{set} = 28.5pA) **d)** example of an (R) herringbone island. (U_{bias} = +1.21V, I_{set} = 28.8 pA)

Another characteristic feature of herringbone structure was the observation that the two protrusions of dark row molecules align with one of the high symmetry directions of the substrate. As the two protrusions resemble the upwards bent benzopyrrole groups of the respective molecule it can be stated that the molecules in the dark row are aligned to the one high symmetry substrate direction along the axis defined by the upwards bent benzopyrrole groups. In principle this means that three different orientations of the molecules in the dark rows are possible according to the three substrate directions. However, chirality of the different orientations has to be regarded. In figure 4.2.6-3 b) the molecules resembled by the dark orange dumbbells align along the high symmetry substrate direction indicated by the green arrow. This is the same situation observed for the molecules in the dark rows of the herringbone structure. In figure 4.2.6-3 b) the two different possibilities are depicted. They can be distinguished by the nominations 'left' (L) and 'right' (R). Concerning an orientation of the dark row molecules vertical to the viewing direction, in the (L) conformation the next higher positioned molecule within one dark row lies always on the left side of the respective neighboring dark row molecule. Therefore an overall left ascending orientation of the predominant direction of the island results. The same principle applies to the (R) conformation. In figure 4.2.6-3 c) and d) two examples for the chiralities are presented. In both scanning tunneling micrographs the dark row molecules are orientated along the same high symmetry substrate direction as indicated by the white arrows.

In consideration of the three different high symmetry substrate directions six different domains are possible in theory. For medium coverage only up to two different orientations were observed in one scanning tunneling micrograph. However, after annealing to 400 K for ten minutes several different herringbone domains were observed. Figure 4.2.6-4 presents an exemplary scanning tunneling micrograph recorded after the annealing step. It should be noted that the coverage before annealing was close to monolayer coverage. After the annealing step only herringbone ordered arrangements were found on the surface. Between the islands no more 2D gas phase but disordered areas were observed. The colored arrows in the center of figure 4.2.6-4 mark the high symmetry directions of the substrate determined by the orientation of the dark row molecules of the herringbone domains. Furthermore the domains are characterized by their (R) or (L) conformation. For each domain one colored letter is assigned which resembles the determined orientation of the respective domain. Basically all the different domains were found and no distinct domain orientation dominated.

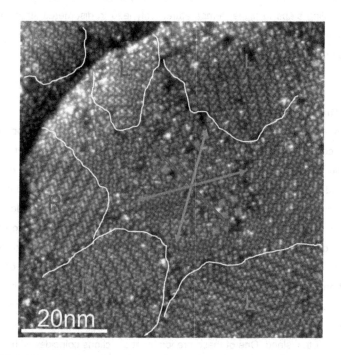

Figure 4.2.6-4 Scanning tunneling micrograph recorded after an annealing step to 400 K for ten minutes. The white lines indicate herringbone domains of different orientation. The colored arrows mark the three different high symmetry directions and therefore the possible dark row molecule orientations. By applying the nomination for the two different chiralities the different domains can be labeled. (U_{bias} = +1.08 V, I_{set} = 28.3 pA)

Motivation for performing the annealing experiments was to examine if the annealing leads to a change in the molecular structure of the Ni-TPBP molecules. For 2HTPP on Cu(111) it was reported that annealing leads to breaking of carbon–hydrogen bonds and forming of new carbon-carbon bonds.[69] This behavior was attributed to the influence of the sterical hindrance of phenyl and pyrrole groups. In STM the change in the molecular constitution was observed in form of a change in the typical 2HTPP contrast. However, for the annealing experiment performed for Ni-TPBP no change in the molecular appearance was observed. After the annealing step the molecules either ordered in the herringbone structure or appeared in form of disordered arrangements of two protrusions resembling single molecules. In addition further annealing steps to higher temperatures only lead to disordered structures and denaturation of Ni-TPBP molecules. Therefore it can be concluded that for Ni-TPBP no structural change, as it was reported for 2HTPP on Cu(111), occurred.

4.2.7. Cross structure

The cross structure was the third and final supramolecular order that was apparent at medium coverage. This assembly consists of square ordered crosses that are build up by four protrusions each. In between the crosses 'vacant areas' were observed. Compared to the hexagonal structure cross structured islands were found more frequently in numbers and in larger sizes. In section 4.2.1 it was presented that the molecules in the structure arrange in a way that they connect neighboring crosses. Thus two different orientations of bridging molecules exist which are rotated by 90° in respect to each other. In principle this perpendicular arrangement reminds one of the herringbone structure. However, in the cross structure the molecules of different orientations are displaced and orientate somewhat towards the center of the crosses that they bridge. Moreover the two different orientations define the two predominant directions of the island. One of these predominant directions coincides with one high symmetry direction of the substrate.

The explanations of the aforementioned observations and especially the arrangement of molecules in the islands are based on the two protrusions per molecule model and the saddle shape intramolecular conformation which was presented in section 4.2.2. However, for the cross structure it was possible to observe four smaller protrusions in the center of the crosses which cannot be explained with the aforementioned models. Considering that for the two protrusion model the upwards bent benzopyrrole groups of saddle shaped Ni-TPBP were assumed to dominate the contrast it is proposed that the observed four smaller protrusions in the center of a cross are caused by four upwards bent benzopyrrole groups of one Ni-TPBP molecule. Under this assumption the tilt angle of the involved benzopyrrole groups can be calculated analog to the calculations for the two protrusions model presented in section 4.2.2. Along these lines a tilt angle of $\alpha_{bpy} = 60°$ was calculated for the four benzopyrrole groups. Figure 4.2.7-1 a) presents a top view of the intramolecular conformation considering the calculated value. Similar as for the saddle shape conformation the parameters for the phenyl legs had to be estimated. In figure 4.2.7-1 b) a side view of the model illustrates that the phenyl legs are tilted towards the surface by 15° while the twist angle equals 0°. Since the metal center atom resembles the bulkiest part of the structure a small tilt angle of the phenyl legs is reasonable to achieve closer position of more groups of the macrocycle to the surface and thus more potential interaction. Furthermore through tilting the phenyl legs towards the surface the steric repulsion of these groups with the benzopyrrole groups is reduced. In consideration of the four small protrusions it is also imaginable that the protrusions are not related to upwards bent benzopyrrole groups but to four upwards bent phenyl legs of a Ni-TPBP molecule. However, calculating the tilt angles for such a conformation from protrusion distances and triangulation yields extreme tilt angles which are not reasonable for a stable intramolecular conformation. In addition the model of four upwards bent benzopyrrole groups is supported by DFT gas phase calculations for the intramolecular conformation of Ni-TPBP. Calculations were performed by the group of Wolfgang Hieringer. Their results are presented in figure 4.2.7-1 c) and d). It has to be stated that this conformation is no real energetic minimum in gas phase and was just possible to calculate by forcing a 4-fold symmetry (C4v). However, this approach is reasonable considering that the substrate has an essential influence on the stability of the conformation and that adsorbed molecules are restrained by interactions with the substrate. The DFT gas phase calculated models yield values of α_{bpy} ≈ 30±5° for the tilt angle of the benzopyrrole groups while the tilt angle of the phenyl legs amounts also 30±5°and the twist angle equals 0°. In comparison to the model presented in figure 4.2.7-1 a) and b), derivations for the DFT gas phase calculated model have to be expected because the energy optimized model does not incorporate the influence of the substrate. The tilt angle of the phenyl legs in the DFT calculated model is certainly unreasonable high because such a conformation would cause a large macrocycle to substrate distance and thus strongly reduced substrate-molecule interactions. As stated before a twist angle of 15° for the phenyl legs, as it is implemented in the model in figure 4.2.7-1 a) and b), is more reasonable to account for the influence of molecule-substrate interactions on the intramolecular conformation. In addition the first model incorporates the protrusion distances in form of the benzopyrrole angles and is thus more related to the situation of an adsorbed molecule on the surface. The smaller tilt angles of the benzopyrrole group of the DFT calculated gas phase model would therefore not fit to the protrusions observed in the center of the crosses. For these reasons the first model will be used to resemble the molecules that are attributed to the four protrusion appearance in the center of the crosses. However, the DFT calculations clearly show that the conformation with all four benzopyrrole groups bent upwards is possible.

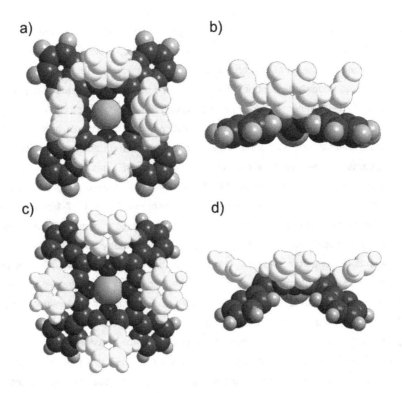

Figure 4.2.7-1 a) proposed model for the intramolecular conformation of Ni-TPBP molecules located in the center of a cross. The model incorporates a tilt angle of 60° for the benzopyrrole groups. The tilt angle of the phenyl legs is 15° to increase molecule-substrate interaction. **b)** sideview of the model in a). **c)** DFT energy minimized gas phase model with fixed C4v symmetry for Ni-TPBP molecules in the center of a cross. The conformation is defined by a tilt angle of $\alpha_{bpy} \approx 30\pm5°$ for the benzopyrrole groups and a tilt angle of 30±5° for the phenyl legs. Calculations were performed by the group of Wolfgang Hieringer. **d)** sideview of the model in c).

By combining the models of the saddle shape conformation and the conformation with four upwards bent benzopyrrole groups the positioning of the molecules and thus the molecule-molecule interactions within the cross structure will be discussed. In figure 4.2.7-2 a high resolution scanning tunneling micrograph of a cross structured island with overlayed molecular models is presented. Interestingly, for this image the appearance of the four protrusions in the center of a cross was more pronounced while the two protrusion contrast attributed to the bridging molecules was less dominant. The bridging molecules are rotated equally counter clockwise in respect to the lattice vectors of the cross structured island. Also the molecules in the center of crosses show similar rotation concerning their orientation of benzoyprrol groups to the lattice vectors. Considering the orientation of molecules one cross is build up by four bridging molecules of saddle shape conformation and one molecule that constitutes four upwards bent benzopyrrole groups. This group of molecules possesses approximately C4 symmetry assuming the nickel atom of the molecule located in the center of the cross as the center of symmetry.

Figure 4.2.7-2 High resolution scanning tunneling micrograph of the cross structure superimposed with molecule models. The red T-shape illustrates the T-type interaction of
phenyl and benzopyrrole groups of bridging molecules. T-type interaction between an
upwards bent benzopyrrole group of a center molecule and the upwards bent benzopyrrole group of a bridging molecule is symbolized by a green T-shape. The dark red
parallel lines symbolize the π-π stacking between benzopyrrole groups of bridging molecules and phenyl legs of center molecules. For marking the π-π stacking of phenyl legs
of bridging molecules and benzopyrrole groups of center molecules purple parallel lines
are given. The π-π stacking between phenyl legs is illustrated by the orange parallel
lines. (U_{bias} = -0.59 V, I_{set} = 29.1 pA)

In order to determine the molecule-molecule interactions possible T-type and π-π
stacking interactions need to be discussed.[11, 43, 63] Since the T-type interactions in
the hexagonal and the herringbone structure were the most reasonable approaches
to explain molecule-molecule interactions, similar interactions are expected to be
prominent in the cross structure as well.

In principle three different possible T-type interactions appear reasonable. One of them is interaction from the para-positioned carbon of a phenyl leg of a bridging molecule to the center of a benzene ring of a downwards bent benzopyrrole group of a bridging molecule. The proposed interaction is illustrated by a red colored T-shape in figure 4.2.7-2. However, the mean center to center distance of interacting groups yields 8.1±0.55 Å. This distance is clearly too large compared to the reported favorable regime of 4.5-7 Å for stable T-type interactions and therefore the proposed interaction is not reasonable.[65-67]

The aforementioned T-type interaction was closely related to the ones proposed for the hexagonal and the herringbone structure. But for the cross structure there is also the possibility of a T-type interaction involving groups of the molecules located in the centers of crosses. The proposed interaction concerns a benzene ring of a benzopyrrole group of a center molecule facing the center of a benzene ring of an upwards bent benzopyrrole group of a bridging molecule. The interaction of interest is marked by a green colored T-shape in figure 4.2.7-2. Considering the position of involved groups to each other the dihedral angle is certainly in the favorable regime of 50-90° reported for T-type interacting aromatic systems.[67] Furthermore the mean center to center distance of 7.1±0.62 Å for the interacting groups is at the limit of the favorable regime. But considering the accuracy of the models and some flexibility of molecules in the structure this distance is still plausible for favorable T-type interactions.

The last discussed T-type interaction, involving groups of a center molecule as well as a bridging molecule, is definitely favored by the observed rotation of bridging molecules and the center molecule in respect to the lattice vectors. In addition this rotation reduces the steric repulsion of upwards bent benzopyrrole groups as well as of phenyl legs of the two molecule types. The rotation is also likely to promote π-π stacking interactions. In order to examine this behavior further possible π-π stacking interactions are discussed.

One possible stacking interaction of π-systems is between phenyl legs of two perpendicular bridging molecules. Orange colored dashed lines indicate the interaction in figure 4.2.7-2. Considering the position of phenyl legs to each other the favorable 'face to face' orientation of interacting aromatic groups is existent. However, the mean center to center distance amounts 5.6±0.55 Å and the center displacement yields 5.4±0.72 Å. These values are far away from the reported optimal distances and thus the proposed interaction can be neglected.

There are also two possible π-π stacking interactions involving groups of a center molecule as well as groups of a bridging molecule. One of them is interaction induced by a benzene ring of an upwards bent benzopyrrole group of a center molecule stacked with a phenyl leg of a bridging molecule. In figure 4.2.7-2 the interaction is marked by purple dashed lines. Concerning that the numerical value of the benzopyrrole tilt angle of the center molecule equals the value of the phenyl twist angle of the bridging molecule excellent conditions for 'face to face' orientation are given. However, for the arrangement of the molecule models in figure 4.2.7-2 the two groups of interest are not aligned exactly parallel. On the other hand it has to be considered that the groups are somewhat flexible and thus, for example, the bridging molecule could be rotated by a small amount to improve the interaction. For the center to center distance a mean value of 3.9±0.51 Å was calculated, while the mean center displacement yields 1.5±0.24 Å. Given the accuracy of measurements and of the models both values are in excellent agreement with the values reported for favorable aromatic group distances.[70] Sherrill et al. state that for a displaced arrangement of two interacting benzene rings the optimal center to center distance equals 3.6 Å while the optimal center displacement is 1.6 Å.[66] Therefore the proposed interaction is very likely to strongly contribute to the molecule-molecule interactions in the cross structure.

Furthermore a π-π stacking interaction between an upwards bent benzopyrrole group of a bridging molecule and a phenyl leg of a center molecule is possible. Two red dashed lines indicate the interaction in figure 4.2.7-2. The mean center to center distance amounts 3.0±0.17 Å and the center displacement constitutes 0.9±0.17 Å. In general these values are smaller than the favorable distances found for aromatic systems. However, also the sterical positioning of the interacting groups has to be regarded. Figure 4.2.7-3 illustrates the positioning of the two interacting groups to each other in form of a sideview of the interacting molecules. In order to achieve 'face to face' orientation either the phenyl leg or the benzopyrrole group has to be tilted more towards the surface. A reasonable approach would combine the two possibilities in a way that the tilt angle of the phenyl leg is increased slightly while the tilt angle of the benzopyrrole group is reduced. Such a reorientation could also increase the center to center and center displacement distances to favorable values. However, steric repulsion induced by neighboring groups might hinder such a reorientation of the ben-

zopyrrole groups. Besides, the benzopyrrole group of the bridging molecule is proposed to be a part of T-type interaction with the benzopyrrole group of the center
molecule. Therefore changes of the tilt angle might negatively influence the overall
stability. For the phenyl leg of the center molecule altering the tilt angle also influ-
ences the overall positioning of groups since the change in conformation increases
the overall distance of the groups to the substrate. Considering the complexity of the
cross structure and the limitations of the molecule models it is difficult to judge the
role of the proposed interaction.

Figure 4.2.7-3 Side view of a center and a bridging molecule emphasizing the π-π stacking
between benzopyrrole group and phenyl leg indicated by the dark red arrow. Ideal 'face
to face' orientation might be achieved through altering the tilt angles of interacting
groups.

All in all for the cross structure a combination of π-π stacking and T-type interactions
accounts for the molecule-molecule interactions. In relation to the dominating interactions it might be possible to explain the rotation of molecules in respect to the lattice
vectors. Most likely the rotation induces a state which balances the attempt of optimizing interactions as well as minimizing steric repulsion. However, different degrees
of rotation were observed for different cross structured islands. It was also possible to
observe islands in which the bridging molecules of the two different predominant directions were rotated irregularly. Also the influence of the substrate has to be considered. Therefore it has to be assumed that different favorable orientations for the
structure exist.

In consideration of the peculiar arrangement of molecules in the cross structure and
the combination of two different intramolecular conformations the unit cell and thus

the molecular density of the structure need to be determined. Figure 4.2.7-4 a) presents a scanning tunneling micrograph of a cross structured island. The distances used to calculate the mean unit cell vectors are indicated by the orange lines while the square unit cell itself is marked by a black square. For the unit cell vector a mean value of $a = 2.52\pm0.12$ nm was calculated while the second unit cell vector amounts $b = 2.63\pm0.12$ nm. The square unit cell is defined by the angle $\alpha = 96\pm2°$. Considering the bridging molecules and the center molecules the unit cell contains three molecules and a molecular density of 0.45 mol./nm² results. If only the bridging molecules are considered the molecular density would yield only 0.30 mol./nm² because in that case the unit cell contains just two molecules per unit cell. This molecular density would be considerably lower than the density of the herringbone structure and even lower than for the hexagonal structure. However, at medium coverage cross structured islands were observed in similar numbers and size as the densely packed herringbone structure whereas the hexagonal was rarely observed. Therefore it would be unlikely that the molecular density of the cross structure is by far the lowest, and thus it can be assumed that the unit cell contains indeed three molecules. Of course also the influence of the overall stability of the island through interactions has to be considered. But the analysis of the interactions as presented above pointed out that the molecule in the center of a cross plays a main role for the interaction of groups. Summing up the discussion it becomes clear that the approach of a molecule located in the center of a cross explains the observations for the cross structure best and is therefore reasonable.

Figure 4.2.7-4 a) scanning tunneling micrograph of the cross structure. The unit cell is indicated by the black square defined a = 2.52±0.12 nm, b = 2.63±0.12 nm and α = 96±2°. (U_{bias} = +1.24 V, I_{set} = 29.1 pA) **b)** STM image presenting three different oriented cross structured islands. The white arrows mark the lattice vector directions which coincide with the high symmetry substrate directions. (U_{bias} = +1.02 V, I_{set} = 29.9 pA) **c)** high resolution image of the cross structure. The black lines mark the lattice vector directions. Dumbbell drawings are used to indicate the counter clockwise rotation of molecules in respect to the lattice vectors. (U_{bias} = +1.20 V, I_{set} = 28.0 pA) **d)** similar high resolution image but in this case the molecules are rotated in a clockwise manner. (U_{bias} = +1.21 V, I_{set} = 28.8 pA)

As a final point for the discussion of cross structure the different orientation of the islands in respect to the substrate will be examined. In section 4.2.1 it was determined that one of the lattice vectors of a cross structured island coincides with one of the high symmetry directions of the substrate. This would mean that only three different orientations of cross structured islands are possible. In figure 4.2.7-4 b) a scanning tunneling micrograph is presented in which three different oriented islands were observed. The white arrows mark the respective lattice vector directions that coincide with the substrate orientations. However, at closer inspection of the cross structure islands it is observed that the bridging molecules as well as the center molecules exhibit some rotation in respect to the lattice vectors of the structure. This behavior was already mentioned for the discussion of interactions at the beginning of this section. In figure 4.2.7-4 c) and d) two more close ups of different cross structured islands are presented. The dark orange dumbbell drawings indicate the arrangement of bridging molecules while the black lines mark the lattice vector directions of the island defined. For the molecules in figure 4.2.7-4 c) the molecules are clearly rotated counter clockwise in respect to the lattice vector while in figure 4.2.7-4 d) the rotation occurs in a clockwise manner. From these images it might be concluded that the cross structure is chiral in two different arrangements. However, it was observed that the degree of rotation of molecules deviates and no general pattern could be distinguished. This behavior was not only apparent comparing different islands but also considering the degree of rotation of molecules in one island. For example in figure 4.2.7-4 c) the molecules located in the vertical direction are rotated more than the molecules located in the horizontal direction. As stated at the end of the discussion for the interactions in the cross structure the rotational movement is most likely connected to balancing optimization of interactions and steric repulsion. Therefore no definite chirality of the cross structure can be concluded.

4.3. High coverage

In order to examine the adsorption behavior of Ni-TPBP on Cu(111) at RT further the coverage was increased to a regime where most parts of the surface were occupied. Figure 4.3-1 presents a scanning tunneling micrograph recorded at coverage close to a monolayer. On the left side a part of a cross structure island and on the right side part of a large herringbone structure was observed. The islands were separated by mostly areas of disordered protrusion arrangements and some stripy features. This image resembles the general situation that was observed on the surface. Interestingly, no more hexagonal ordered islands were observed while the cross and herringbone islands were still found. In general the herringbone structure was the most dominating arrangement concerning the size and number of islands.

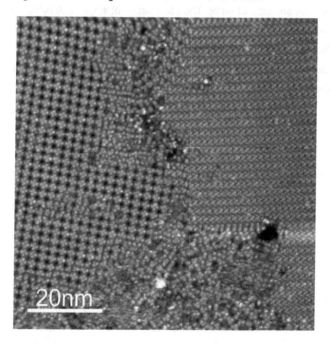

Figure 4.3-1 Scanning tunneling micrograph recorded at high coverage of Ni-TPBP on Cu(111) at RT. On the left side part of a cross structured island was visible, while on the right side part of a large herringbone island was found. The islands were separated by areas of disordered arrangement of protrusions attributed to single molecules. ($U_{bi\text{-}as}$ = +1.03 V, I_{set} = 28.4 pA)

Considering comparable interactions of molecules in the different structures, generally, the structure with the highest molecular is the most favorable at high coverages.
This conclusion is in good agreement with the observations made for high coverage
as presented in figure 4.3-1. However, the numerical value of the molecular density
of the hexagonal structure (0.48 mol./nm^2) is higher than the value calculated for the
cross structure (0.45 mol./nm^2). But as stated before the hexagonal structure was not
observed while the cross structure was still found on the surface. Therefore it can be
concluded that the molecule-molecule interactions in the cross structure are stronger
than in the hexagonal structure. Thus the cross structure presents a more favorable
arrangement and prevails over the hexagonal structure.

The disordered areas in between the islands are attributed to single molecules that
arrange without any distinct order. The observation of single molecules is most likely
due to a reduction of the mean free path of diffusion for the molecules of the 2D gas
phase. The possibilities of diffusion get limited through the high coverage of the surface which is also apparent in form of a reduced 2D gas phase indicated by less observed stripy features.

5. Conclusion and Outlook

In the following the adsorption behavior of Ni-TPBP on Cu(111) investigated with STM under UHV at RT will be summarized briefly.

Starting with low coverage individual stationary molecules are only found arranged in chains agglomerated at the step edges of the surface.[52-53] While on the terraces Ni-TPBP molecules are probably freely diffusing on the surface and cannot be imaged as individual entities.[51]

At medium coverage three different coexisting supramolecular arrangements, namely a hexagonal, a cross and a herringbone structure of Ni-TPBP on Cu(111) are observed. The rarely found hexagonal structure is characterized by a hexagonal arrangement of bright protrusions while the herringbone structure shows alternating dark and bright rows. The cross structure resembles a square arrangement of crosses that are built up by four protrusions each. In addition to the stationary molecules in the supramolecular structures also attaching and detaching molecules at the island boundaries are observed. Furthermore there is also evidence for freely diffusing molecules in a 2D gas phase. From the attachment/detachment of molecules at the boundaries of the islands one Ni-TPBP molecule could be identified as two protrusions in STM.

With this approach it was possible to determine the molecular arrangement in the structures as well as the intramolecular conformation which leads to the observed STM contrast. The intramolecular conformation of Ni-TPBP on Cu(111) resembles a saddle-shape conformation similar to the ones found for other porphyrin systems.[42] Therefore the 'two protrusion' contrast observed in STM is caused by two benzene rings of upwards bent opposite benzopyrrole groups. The dark center between the protrusions can be attributed to the nickel center.[56-57] Through triangulation of the two protrusions a tilt angle $\alpha_{bpy} = 45°$ for the benzopyrrole groups in the saddle shape conformation was determined. Since the STM measurements did not allow for the determination of the phenyl substituents conformation a twist angle $\alpha_{ph} = 60°$ of the latter was adapted from a comparable porphyrin system. However, the assumed saddle shape conformation could not explain an observation of four smaller protrusions in the center of a cross in the cross structured islands. This feature was attributed to a different intramolecular conformation in which all four benzopyrrole groups are bent upwards by 60° and the phenyl legs are inclined towards the surface by 15°. Based on DFT gas phase calculations, conducted by the group of Wolfgang Hieringer, it was concluded that Ni-TPBP exhibits a global minimum in the saddle shape conformation supporting the assumption derived from the STM data. For the four protrusion structure calculation of the intramolecular conformation was possible by forcing C4v symmetry. Considering that the DFT gas phase calculations did not account for the influence of the substrate as well as intermolecular interactions the agreement of models derived from STM data and energy optimized models is very good.

Detailed analysis of the structures yields that in the hexagonal and in the herringbone structure T-type interactions are the dominating molecule-molecule interactions. However, in the herringbone structure more regular ordering of the molecules as well as shorter distances of interacting groups result in a higher molecular density and an overall more favorable arrangement compared to the hexagonal structure. The cross structure resembles a very different situation since it combines two different intramolecular conformations. The bridging and center molecules arranged in a way that favorable conditions for both T-type and π-π stacking were accomplished causing a stable arrangement despite the lower molecular density.

The stability of herringbone and cross structure as well as the comparably unfavorable situation of the hexagonal structure is also reflected in the situation for coverages close to a monolayer, i.e., no hexagonal structure is found, while herringbone and cross structure are evenly distributed. After moderate annealing it is found that only the herringbone structure remains which is in good agreement with the highest molecular density and the proposed stability of the arrangement.

All in all the results in the work at hand demonstrate that TBP's present an interesting class of molecules with very unusual adsorption behavior, i.e. exhibiting a peculiar polymorphism. In particular for Ni-TPBP on Cu(111) the observation of three different coexisting supramolecular arrangements and the discovery of two different intramolecular conformations within one arrangement are striking and of interest considering fabrication of tailor-made functional molecular architectures.

References

[1]   J. V. Barth, *Annual review of physical chemistry* **2007**, *58*, 375-407.

[2]   J. V. Barth, J. Weckesser, C. Cai, P. Günter, L. Bürgi, O. Jeandupeux, K. Kern, *Angew. Chem. Int. Ed.* **2000**, *39*, 1230-1234.

[3]   S. De Feyter, F. C. De Schryver, *Chemical Society Reviews* **2003**, *32*, 139-150.

[4]   S. R. Forrest, *Chem. Rev.* **1997**, *97*, 1793-1896.

[5]   B. A. Hermann, L. J. Scherer, C. E. Housecroft, E. C. Constable, *Advanced Functional Materials* **2006**, *16*, 221-235.

[6]   J. V. Barth, *Surface Science* **2009**, *603*, 1533-1541.

[7]   Y. Bai, F. Buchner, I. Kellner, M. Schmid, F. Vollnhals, H.-P. Steinrück, H. Marbach, J. Michael Gottfried, *New Journal of Physics* **2009**, *11*, 125004.

[8]   W. Auwärter, A. Weber-Bargioni, S. Brink, A. Riemann, A. Schiffrin, M. Ruben, J. V. Barth, *Chemphyschem : a European journal of chemical physics and physical chemistry* **2007**, *8*, 250-254.

[9]   A. Weber-Bargioni, W. Auwärter, F. Klappenberger, J. Reichert, S. Lefrancois, T. Strunskus, C. Woll, A. Schiffrin, Y. Pennec, J. V. Barth, *Chemphyschem : a European journal of chemical physics and physical chemistry* **2008**, *9*, 89-94.

[10]   C. M. Carvalho, T. J. Brocksom, K. T. de Oliveira, *Chemical Society reviews* **2013**, *42*, 3302-3317.

[11]   J. Brede, M. Linares, S. Kuck, J. Schwöbel, A. Scarfato, S. H. Chang, G. Hoffmann, R. Wiesendanger, R. Lensen, P. H. Kouwer, J. Hoogboom, A. E. Rowan, M. Broring, M. Funk, S. Stafstrom, F. Zerbetto, R. Lazzaroni, *Nanotechnology* **2009**, *20*, 275602.

[12]   F. Buchner, K. Comanici, N. Jux, H. P. Steinrück, H. Marbach, *J. Phys. Chem. C* **2007**, *111*, 13531-13538.

[13]   G. Rojas, X. Chen, C. Bravo, J.-H. Kim, J.-S. Kim, J. Xiao, P. Dowben, Y. Gao, X. C. Zeng, W. Choe, A. Enders, *J. Phys. Chem. C* **2010**, *114*, 9408-9415.

[14]   Y. Matsuzawa, K. Ichimura, K. Kudo, *Inorganica Chimica Acta* **1998**, *277*, 151-156.

[15]   A. V. Zamyatin, Bowling Green State University **2006**.

84   References

[16]   K. M. Kadish, K. M. Smith, R. Guliard, *The Porphyrin Handbook: Synthesis and organic chemistry, Vol. 1*, Academic Press, San Diego, **2000**.

[17]   R. Hattori, C. Shim, S. Lee, M. Tazoe, T. Nakashima, S. Aramaki, A. Ohno, Y. Sakai, *IDW* **2006**.

[18]   J. Mack, M. Bunya, Y. Shimizu, H. Uoyama, N. Komobuchi, T. Okujima, H. Uno, S. Ito, M. J. Stillman, N. Ono, N. Kobayashi, *Chemistry* **2008**, *14*, 5001-5020.

[19]   P. B. Shea, A. R. Johnson, N. Ono, J. Kanicki, *IEEE Transactions on electronic devices* **2005**, *52*, 1497-1503.

[20]   G. Binnig, H. Rohrer, C. Gerber, E. Weibel, *Physical Review Letters* **1982**, *49*, 57-61.

[21]   G. Binnig, H. Rohrer, C. Gerber, E. Weibel, *Physical Review Letters* **1983**, *50*, 120-123.

[22]   L. d. Broglie, *Annales de Physique* **1925**, *10*.

[23]   G. Wedler, *Lehrbuch der Physikalischen Chemie*, 4 ed., Wiley-VCH, Weinheim, **1997**.

[24]   P. W. Atkins, *Physikalische Chemie*, 1 ed., VCH, Weinheim, Basel, Cambridge, New York, **1987**.

[25]   J. C. Chen, *Introduction to Scanning Tunneling Microscopy*, Oxford University Press, New York, **1993**.

[26]   M.Schmid, *http://upload.wikimedia.org/wikipedia/commons/f/f9/ScanningTunnelingMicroscope_schematic.png*, Accessed March 2014.

[27]   K. Comanici, Ph.D. thesis, FAU Erlangen-Nürnberg **2007**.

[28]   J. Tersoff, D. R. Hamann, *Physical Review Letters* **1983**, *50*, 1998-2001.

[29]   J. Tersoff, D. R. Hamann, *Physical Review B* **1985**, *31*, 805-813.

[30]   A. Baratoff, *Physica* **1984**, *127B*, 143-150.

[31]   S. Ohnishi, M. Tsukada, *Solid State Communications* **1989**, *71*, 391-394.

[32]   J. C. Chen, in *Topical Conference on Nanometer Scale Properties of Surfaces and Interfaces, Vol. 9*, AVS, Boston, Massachusetts (USA), **1991**, pp. 44-50.

[33]   B. M. Tissue, *http://www.files.chem.vt.edu/chem-ed/ms/quadrupo.html*, Accessed March 2014.

[34]   H. Wiberg, *Anorganische Chemie*.

[35]   A. Görling, *Script to the lecture 'Theorie periodischer Systeme'*, **2011**.

[36] F. Buchner, Ph. D. thesis, FAU Erlangen-Nürnberg 2010.

[37] M.-C. Marinica, C. Barreteau, M.-C. Desjonquères, D. Spanjaard, *Physical Review B* 2004, *70*.

[38] M.-C. Marinica, C. Barreteau, D. Spanjaard, M.-C. Desjonquères, *Physical Review B* 2005, *72*.

[39] P. Schaeffer, R. Ocampo, H. J. Callot, P. Albrecht, *Nature* 1993, *364*, 133-136.

[40] K. Peter, C. Vollhardt, N. E. Schore, *Organische Chemie*, 4 ed., Wiley-VCH, Weinheim, 2005.

[41] F. Buchner, E. Zillner, M. Röckert, S. Gläßel, H. P. Steinrück, H. Marbach, *Chem. Eur. J.* 2011, *17*, 10226-10229.

[42] K. Diller, F. Klappenberger, M. Marschall, K. Hermann, A. Nefedov, C. Woll, J. V. Barth, *J Chem Phys* 2012, *136*, 014705.

[43] F. Buchner, I. Kellner, W. Hieringer, A. Gorling, H. P. Steinruck, H. Marbach, *Physical chemistry chemical physics : PCCP* 2010, *12*, 13082-13090.

[44] S. Ditze, Ph. D. thesis, FAU Erlangen-Nürnberg 2014.

[45] F. Buchner, K.-G. Warnick, T. Wölfle, A. Görling, H. P. Steinrück, W. Hieringer, H. Marbach, *J. Phys. Chem. C* 2009, *113*, 16450-16457.

[46] F. Buchner, J. Xiao, E. Zillner, M. Chen, M. Röckert, S. Ditze, M. Stark, H.-P. Steinrück, J. M. Gottfried, H. Marbach, *The Journal of Physical Chemistry C* 2011, *115*, 24172-24177.

[47] S. Ditze, M. Stark, M. Drost, F. Buchner, H. P. Steinruck, H. Marbach, *Angewandte Chemie* 2012, *51*, 10898-10901.

[48] I. RHK Technology, *UHV 300 Variable Temperature Ultrahigh Vacuum Scanning Tunneling Microscope, User's Guide*, Troy, Michigan (USA), 1998.

[49] I. Horcas et al., *Review of Scientific Instruments, Vol. 78*, 2007.

[50] J. V. Barth, *Surface Science Reports* 2000, *40*, 75-149.

[51] H. Yanagi, H. Mukai, K. Ikuta, T. Shibutani, T. Kamikado, S. Yokoyama, S. Mashiko, *Nano Letters* 2002, *2*, 601-604.

[52] F. Besenbacher, *Rep. Prog. Phys.* 1996, *59*, 1737-1802.

[53] J. J. de Miguel, R. Miranda, *J. Phys.: Condens. Matter* 2002, *14*, 1063-1097.

[54] D. van Vorden, M. Lange, M. Schmuck, J. Schaffert, M. C. Cottin, C. A. Bobisch, R. Moller, *J Chem Phys* 2013, *138*, 211102.

[55] B. W. Heinrich, G. Ahmadi, V. L. Müller, L. Braun, J. I. Pascual, K. J. Franke,
      Nano Lett **2013**, *13*, 4840-4843.

[56] X. Lu, K. W. Hipps, *J. Phys. CHEM. B* **1997**, *101*, 5391-5396.

[57] L. Scudiero, D. E. Barlow, K. W. Hipps, *J. Phys. CHEM. B* **2000**, *104*, 11899-
      11905.

[58] X. Lu, K. W. Hipps, X. D. Wang, U. Mazur, *J. Am. Chem. Soc.* **1996**, *118*,
      7197-7202.

[59] S. Ditze, M. Stark, F. Buchner, A. Aichert, N. Jux, N. Luckas, A. Gorling, W.
      Hieringer, J. Hornegger, H. P. Steinruck, H. Marbach, *Journal of the American
      Chemical Society* **2014**, *136*, 1609-1616.

[60] S. Berner, M. Brunner, L. Ramoino, H. Suzuki, H.-J. Güntherodt, T. A. Jung,
      Chemical Physics Letters **2001**, *348*, 171-181.

[61] M. Roeckert, Degree thesis, FAU Erlangen-Nürnberg **2010**.

[62] E. Zillner, Master thesis, FAU Erlangen-Nürnberg **2009**.

[63] M. Stark, S. Ditze, M. Drost, F. Buchner, H. P. Steinrück, H. Marbach,
      Langmuir : the ACS journal of surfaces and colloids **2013**, *29*, 4104-4110.

[64] S. Grimme, *Angewandte Chemie* **2008**, *47*, 3430-3434.

[65] P. C. Jha, Z. Rinkevicius, H. Agren, P. Seal, S. Chakrabarti, *Physical
      chemistry chemical physics : PCCP* **2008**, *10*, 2581-2583.

[66] M. O. Sinnokrot, C. D. Sherrill, *J. Phys. Chem. A* **2006**, *110*, 10656-10668.

[67] S. K. Burley, G. A. Petsko, *Science* **1985**, *229*, 23-28.

[68] E. Arunan, H. S. Gutowsky, *The Journal of Chemical Physics* **1993**, *98*, 4294.

[69] J. Xiao, S. Ditze, M. Chen, F. Buchner, M. Stark, M. Drost, H.-P. Steinrück, J.
      M. Gottfried, H. Marbach, *The Journal of Physical Chemistry C* **2012**, *116*,
      12275-12282.

[70] T. Dahl, *Acta Chem. Scand.* **1994**, *48*, 95-106.

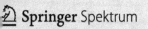

Printed in the United States
By Bookmasters